CONGRESSIONAL POLICIES, PRACTICES AND PROCEDURES

KEY CONGRESSIONAL REPORTS FOR MAY 2019

TECHNOLOGY

Congressional Policies, Practices and Procedures

Additional books and e-books in this series can be found on Nova's website under the Series tab.

CONGRESSIONAL POLICIES, PRACTICES AND PROCEDURES

KEY CONGRESSIONAL REPORTS FOR MAY 2019

TECHNOLOGY

ASTON DUNKLEY
EDITOR

snova
New York

Library of Congress Cataloging-in-Publication Data

ISBN: 978-1-53616-335-3

Published by Nova Science Publishers, Inc. † New York

CONTENTS

PREFACE

This book is a comprehensive compilation of all reports, testimony, correspondence and other publications issued by the Congressional Research Service on International Relations during the month of May.

Chapter 1 - Technological convergence, in general, refers to the trend or phenomenon where two or more independent technologies integrate and form a new outcome. One example is the smartphone. A smartphone integrated several independent technologies—such as telephone, computer, camera, music player, television (TV), and geolocating and navigation tool—into a single device. The smartphone has become its own, identifiable category of technology, establishing a $350 billion industry. Of the three closely associated convergences—technological convergence, media convergence, and network convergence—consumers most often directly engage with technological convergence. Technological convergent devices share three key characteristics. First, converged devices can execute multiple functions to serve blended purpose. Second, converged devices can collect and use data in various formats and employ machine learning techniques to deliver enhanced user experience. Third, converged devices are connected to a network directly and/or are interconnected with other devices to offer ubiquitous access to users. Technological convergence may present a range of issues where Congress may take legislative and/or oversight actions. Three selected issue areas associated with technological convergence are regulatory jurisdiction, digital privacy, and data security. First, merging and integrating multiple technologies from distinct functional categories into one converged technology may pose challenges to defining regulatory policies and responsibilities. Determining oversight jurisdictions and regulatory authorities for converged technologies can become unclear as the boundaries that once separated single-function technologies blend together. A challenge for Congress may be in delineating which government agency has jurisdiction over various converged technologies. Defining policies that regulate technological convergence industry may not be simple or straightforward. This may further complicate how Congress oversees government agencies and converged industries due to blending boundaries of existing categories. Second, converged technologies collect and use personal and machine data which may raise digital privacy concerns for consumers. Data collection and usage are tied to digital privacy issues because a piece or aggregation of information could identify an individual or reveal patterns in one's activities. Converged or smart technologies leverage large volumes of data to try to improve the user experience by generating more tailored and anticipatory results. However, such data can potentially identify, locate, track, and monitor an individual without the person's knowledge. Such data can also potentially be sold to third-party entities without an individual's awareness. As the use of converged technologies continues to propagate, digital privacy issues will likely remain central. Third, data security concerns are often associated with smart devices' convenient ubiquitous features that may double as vulnerabilities exploited by malicious actors. Data security, a component of cybersecurity, protects data from unauthorized access and use. Along with digital privacy, data security is a pertinent issue to technological convergence. As converged devices generate and consume large volumes of data, multiple data security concerns have emerged: potentially increased number of access points susceptible to cyberattacks, linkage to physical security, and theft of data. Relatively few policies are in place for specifically overseeing technological convergence, and current federal data protection laws have varied privacy and data security provisions for different types of personal data. To address regulatory, digital privacy, and data security issues,

Congress may consider the role of the federal government in an environment where technological evolution changes quickly and continues to disrupt existing regulatory frameworks. Regulating technological convergence may entail policies for jurisdictional deconfliction, harmonization, and expansion to address blended or new categories of technology. One approach could be for Congress to define the role of federal government oversight of digital privacy and data security by introducing new legislation that comprehensively addresses digital privacy and data security issues or by expanding the current authorities of federal agencies. When considering new legislation or expanding the authorities of federal agencies, three potential policy decisions are (1) whether data privacy and data security should be addressed together or separately, (2) whether various types of personal data should be treated equally or differently, and (3) which agencies should be responsible for implementing any new laws.

Chapter 2 - As the global internet develops and evolves, digital trade has become more prominent on the global trade and economic policy agenda. The economic impact of the internet was estimated to be $4.2 trillion in 2016, making it the equivalent of the fifth-largest national economy. The digital economy accounted for 6.9% of current-dollar gross U.S. domestic product (GDP) in 2017. Digital trade has been growing faster than traditional trade in goods and services. Congress has an important role to play in shaping global digital trade policy, from oversight of agencies charged with regulating cross-border data flows to shaping and considering legislation implementing new trade rules and disciplines through trade negotiations. Congress also works with the executive branch to identify the right balance between digital trade and other policy objectives, including privacy and national security. Digital trade includes end-products, such as downloaded movies, and products and services that rely on or facilitate digital trade, such as productivity-enhancing tools like cloud data storage and email. In 2017, U.S. exports of information and communications technology-enabled services (excluding digital goods) were an estimated $439 billion. Digital trade is growing on a global basis, contributing more to global domestic product (GDP) than financial or merchandise flows. The increase in digital trade raises new challenges in U.S. trade policy, including how to best address new and emerging trade barriers. As with traditional trade barriers, digital trade constraints can be classified as tariff or nontariff barriers. In addition to high tariffs, barriers to digital trade may include localization requirements, cross border data flow limitations, intellectual property rights (IPR) infringement, forced technology transfer, web filtering, economic espionage, and cybercrime exposure or state-directed theft of trade secrets. China's policies, in particular, such as those on internet sovereignty and cybersecurity, pose challenges for U.S. companies. Digital trade issues often overlap and cut across policy areas, such as IPR and national security; this raises questions for Congress as it weighs different policy objectives. The Organisation for Economic Co-operation and Development (OECD) points out three potentially conflicting policy goals in the internet economy: (1) enabling the internet; (2) boosting or preserving competition within and outside the internet; and (3) protecting privacy and consumers, more generally. While no multilateral agreement on digital trade exists in the World Trade Organization (WTO), other WTO agreements cover some aspects of digital trade. Recent bilateral and plurilateral agreements have begun to address digital trade rules and barriers more explicitly. For example, the proposed U.S.-Mexico-Canada Agreement (USMCA) and ongoing plurilateral discussions in the WTO on a potential e-commerce agreement could address digital trade barriers to varying degrees. Digital trade is also being discussed in a variety of international forums, providing the United States with multiple opportunities to engage in and shape global norms. With workers in the high-tech sector in every U.S. state and congressional district, and over two-thirds of U.S. jobs requiring digital skills, Congress has an interest in ensuring and developing the global rules and norms of the internet economy in line with U.S. laws and norms, and in establishing a U.S. trade policy on digital trade that advances U.S. interests.

Chapter 3 - President Trump's budget request for FY2020 includes approximately $134.1 billion for research and development (R&D). Several FY2019 appropriations bills had not been enacted at the time the President's FY2020 budget was prepared; therefore, the President's budget included the FY2018 actual funding levels, 2019 annualized continuing resolution (CR) levels, and the FY2020 request levels. On February 15, 2019, Congress enacted the Consolidated Appropriations Act, 2019 (P.L. 116-6). This act included each of the remaining appropriations acts, completing the FY2019 appropriations process. The act also rendered the CR levels identified in the budget no longer relevant, though for some agencies the exact amount of R&D funding in the act remained uncertain. The analysis of government-wide R&D funding in this chapter compares the President's request for FY2020 to the FY2018 level. For agencies for which the FY2019 R&D funding levels are known, individual agency analyses in this chapter compare the FY2020 request to FY2019 enacted levels. For agencies for which the FY2019 R&D funding levels remain unknown, individual agency analyses in this chapter compare the FY2020 request to FY2018 actual levels. As of the date of this chapter, the House had not completed action on any of the 12 regular appropriations bills for FY2020; nor had the Senate. In FY2018, OMB adopted a change to the definition of development, applying a more narrow treatment it describes as "experimental development." This change was intended to harmonize the reporting of U.S. R&D funding data with the approach used by other nations. The new definition is used in this chapter. Under the new definition of R&D (applied to both FY2018 and FY2020 figures), President Trump is requesting approximately $134.1 billion for R&D for FY2020, a decrease of $1.7 billion (1.2%) from the FY2018 level. Adjusted for inflation, the President's FY2020 R&D request represents a decrease of 5.1% below the FY2018 level. Funding for R&D is concentrated in a few departments and agencies. In FY2018, eight federal agencies received 96.3% of total federal R&D funding, with the Department of Defense (DOD, 38.6%) and the Department of Health and Human Services (HHS, 27.2%) combined accounting for nearly two-thirds of all federal R&D funding. The same eight agencies account for 97.2% of the FY2020 request, with DOD accounting for 44.3% and HHS for 25.1%. Under the President's FY2020 budget request, most federal agencies would see their R&D funding decline. The primary exception is the Department of Defense. DOD's requested R&D funding for FY2020 is $7.1 billion

(13.5%) above the FY2018 level. The Departments of Transportation and Veterans Affairs would see small increases in R&D funding. Among the agencies with the largest proposed reductions in R&D funding in the FY2020 budget compared to the FY2018 actual levels are the Department of Energy ($2.8 billion, 15.8%), the National Science Foundation ($567 million, 9.0%), and National Aeronautics and Space Administration ($475 million, 4.0%). The President's FY2020 budget request would reduce funding for basic research by $1.5 billion (4.0%), applied research by $4.3 billion (10.5%), and facilities and equipment by $0.5 billion (12.8%), while increasing funding for development by $4.5 billion (8.3%). President Trump's FY2020 budget is largely silent on funding levels for multiagency R&D initiatives. However, some activities supporting these initiatives are discussed in agency budget justifications and are reported in the agency analyses in this chapter. The request represents the President's R&D priorities. Congress may opt to agree with none, part, or all of the request, and it may express different priorities through the appropriations process. In recent years, Congress has completed the annual appropriations process after the start of the fiscal year. Completing the process after the start of the fiscal year and the accompanying use of continuing resolutions can affect agencies' execution of their R&D budgets, including the delay or cancellation of planned R&D activities and the acquisition of R&D-related equipment.

Chapter 4 - U.S. airborne electronic warfare (EW) programs involve developing and procuring EW aircraft and EW systems that are mounted on U.S. aircraft. The President's FY2020 budget request for the Department of Defense (DOD) proposes funding for a number of airborne EW programs. *The Role of Airborne EW in Modern Warfare.* EW is a component of modern warfare, particularly in response to threats posed by potential adversaries such as Russia or China. EW refers to operations that use the electromagnetic spectrum (i.e., the "airwaves") to detect, listen to, jam, and deceive (or "spoof") enemy radars, radio communication systems, data links, and other electronic systems. EW also refers to operations that defend against enemy attempts to do the same. The shift in the international security environment from the post-Cold War era to an era of renewed great power competition has led to an increased focus on EW in U.S. defense planning and programming, particularly aspects of EW related to high-end warfare. *U.S. Airborne Electronic Attack Capabilities.* Airborne EW capabilities are a component of U.S. military airpower. Although dedicated U.S. EW aircraft are relatively few in number compared with U.S. fighters, strike fighters, and attack aircraft, they play a role in helping to ensure the combat survivability and effectiveness of other aircraft and friendly forces on the ground. DOD's three primary manned EW electronic attack aircraft are the Navy EA-18G Growler, the Air Force EC-130H Compass Call, and the Air Force EC-37B Compass Call Re-Host. A fourth manned aircraft—the F-35 Joint Strike Fighter—has extensive, integrated EW capabilities. DOD's primary airborne electronic attack payloads include the AN/ALQ-99 electronic attack suite, the Next Generation Jammer, and the Miniature Air Launched Decoy-Jammer. *EW Oversight Issues for Congress.* Congress has continually shown interest in EW, and the decisions it makes regarding EW could affect future U.S. military capabilities and funding requirements. In particular, EW programs pose several potential oversight issues for Congress

- Whether DOD is prioritizing appropriately airborne EW programs in its planning and budgeting relative to other U.S. military EW programs (such as those for U.S. ground forces or Navy surface ships) and to other DOD non-EW priorities.
- Whether DOD's proposed mix of airborne EW capabilities and investments is appropriate.
- The evolution of technology and how new technologies can be employed for EW operations.
- The Air Force's planned rate for procuring EC-37Bs and replacing EC-130Hs.

Chapter 5 - Electronic forms of payment have become increasingly available, convenient, and cost efficient due to technological advances in digitization and data processing. Anecdotal reporting and certain analyses suggest that businesses and consumers are increasingly eschewing cash payments in favor of electronic payment methods. Such trends have led analysts and policymakers to examine the possibility that the use and acceptance of cash will significantly decline in coming years and to consider the effects of such an evolution. Cash is still a common and widely accepted payment system in the United States. Cash's advantages include its simplicity and robustness as a payment system that requires no ancillary technologies. In addition, it provides privacy in transactions and protection from cyber threats or financial institution failures. However, using cash involves costs to businesses and consumers who pay fees to obtain, manage, and protect cash and exposes its users to loss through misplacement, theft, or accidental destruction of physical currency. Cash also concurrently generates government revenues through "profits" earned by producing it and by acting as interest-free liabilities to the Federal Reserve (in contrast to reserve balances on which the Federal Reserve pays interest), while reducing government revenues by facilitating some tax avoidance. The relative advantages and costs of various payment methods will largely determine whether and to what degree electronic payment systems will displace cash. Traditional noncash payment systems (such as credit and debit cards and interbank clearing systems) involving intermediaries such as banks and central banks address some of the shortcomings of cash payments. These systems can execute payments over physical distance, allow businesses and consumers to avoid some of the costs and risks of using cash, and are run by generally trusted and closely regulated intermediaries. However, the maintenance and operation of legacy noncash systems involve their own costs, and the intermediaries charge fees to recoup those costs and earn profits. The time it takes to finalize certain transactions—including crediting customer accounts for check or electronic deposits—can lead to consumers incurring additional costs. In addition, these systems involve cybersecurity risks and generally require customers to divulge their private personal information to gain system access, which raises privacy concerns. To date, the migration away from cash has largely been in favor of traditional noncash payment systems; however, some observers predict new alternative systems will play a larger role in the future. Such alternative systems aim to address some of the inefficiencies and risks of traditional noncash systems, but face obstacles to achieving that aim and involve costs of their own. Private systems using distributed ledger technology, such as cryptocurrencies, may not serve the main functions of money well and face challenges to widespread acceptance and technological scalability. These systems also raise concerns among certain observers related to whether these systems could facilitate crime, provide inadequate protections to consumers, and may adversely affect governments' ability to implement or transmit monetary policy. The potential for increased payment efficiency from these systems is promising enough that certain central banks have investigated the possibility of issuing government-backed, electronic-only currencies—called central bank digital currencies (CBDCs)—in such a way that the benefits of certain alternative payment systems could be realized with appropriately mitigated risk. How CBDCs would be created and function are still matters of speculation at this time, and the possibility of their introduction raises questions about the appropriate role of a central bank in the financial system and the economy. If the relative benefits and costs of cash and the various other payment methods evolve in such a way that cash is significantly displaced as a commonly accepted form of payment, that evolution could have a number of effects, both positive and negative, on the economy and society. Proponents of reducing cash usage (or even eliminating it all together and becoming a *cashless society*) argue that doing so will generate important benefits, including potentially improved efficiency of the payment system, a reduction of crime, and less constrained monetary policy. Proponents of maintaining cash as a payment option argue that significant reductions in cash usage and acceptance would further marginalize people with limited access to the financial system, increase the financial system's vulnerability to cyberattack, and reduce personal privacy. Based on their assessment of the magnitude of these benefits and costs and the likelihood that market forces will displace cash as a payment system, policymakers may choose to encourage or discourage this trend.

In: Key Congressional Reports for May 2019 ISBN: 978-1-53616-335-3
Editor: Aston Dunkley

Chapter 1

TECHNOLOGICAL CONVERGENCE: REGULATORY, DIGITAL PRIVACY, AND DATA SECURITY ISSUES*

Suzy E. Park

ABSTRACT

Technological convergence, in general, refers to the trend or phenomenon where two or more independent technologies integrate and form a new outcome. One example is the smartphone. A smartphone integrated several independent technologies—such as telephone, computer, camera, music player, television (TV), and geolocating and navigation tool—into a single device. The smartphone has become its own, identifiable category of technology, establishing a $350 billion industry.

Of the three closely associated convergences—technological convergence, media convergence, and network convergence—consumers most often directly engage with technological convergence. Technological convergent devices share three key characteristics. First, converged devices can execute multiple functions to serve blended

* This is an edited, reformatted and augmented version of a Congressional Research Service publication R45746, prepared for Members and Committees of Congress dated May 30, 2019.

purpose. Second, converged devices can collect and use data in various formats and employ machine learning techniques to deliver enhanced user experience. Third, converged devices are connected to a network directly and/or are interconnected with other devices to offer ubiquitous access to users.

Technological convergence may present a range of issues where Congress may take legislative and/or oversight actions. Three selected issue areas associated with technological convergence are regulatory jurisdiction, digital privacy, and data security. First, merging and integrating multiple technologies from distinct functional categories into one converged technology may pose challenges to defining regulatory policies and responsibilities. Determining oversight jurisdictions and regulatory authorities for converged technologies can become unclear as the boundaries that once separated single-function technologies blend together. A challenge for Congress may be in delineating which government agency has jurisdiction over various converged technologies. Defining policies that regulate technological convergence industry may not be simple or straightforward. This may further complicate how Congress oversees government agencies and converged industries due to blending boundaries of existing categories.

Second, converged technologies collect and use personal and machine data which may raise digital privacy concerns for consumers. Data collection and usage are tied to digital privacy issues because a piece or aggregation of information could identify an individual or reveal patterns in one's activities. Converged or smart technologies leverage large volumes of data to try to improve the user experience by generating more tailored and anticipatory results. However, such data can potentially identify, locate, track, and monitor an individual without the person's knowledge. Such data can also potentially be sold to third-party entities without an individual's awareness. As the use of converged technologies continues to propagate, digital privacy issues will likely remain central.

Third, data security concerns are often associated with smart devices' convenient ubiquitous features that may double as vulnerabilities exploited by malicious actors. Data security, a component of cybersecurity, protects data from unauthorized access and use. Along with digital privacy, data security is a pertinent issue to technological convergence. As converged devices generate and consume large volumes of data, multiple data security concerns have emerged: potentially increased number of access points susceptible to cyberattacks, linkage to physical security, and theft of data.

Relatively few policies are in placc for specifically overseeing technological convergence, and current federal data protection laws have varied privacy and data security provisions for different types of personal data. To address regulatory, digital privacy, and data security issues, Congress may consider the role of the federal government in an environment where technological evolution changes quickly and continues to disrupt existing regulatory frameworks. Regulating technological convergence may entail policies for jurisdictional deconfliction, harmonization, and expansion to address blended or new categories of technology. One approach could be for Congress to define the role of federal government oversight of digital privacy and data security by introducing new legislation that comprehensively addresses digital privacy and data security issues or by expanding the current authorities of federal agencies. When considering new legislation or expanding the authorities of federal agencies, three potential policy decisions are (1) whether data privacy and data security should be addressed together or separately, (2) whether various types of personal data should be treated equally or differently, and (3) which agencies should be responsible for implementing any new laws.

INTRODUCTION

Technological convergence, in general, refers to the trend or phenomenon where two or more independent technologies integrate and form a new outcome. One example is the smartphone. A smartphone integrates several independent technologies—such as telephone, computer, camera, music player, television (TV), and geolocating and navigation tool—into a single device. The smartphone has become its own, identifiable category of technology. Currently, over 35% of the global population are smartphone users and over 3 billion active devices are in circulation.[1] In the United States, about 80% of the U.S. population are smartphone users, and over 280 million active devices are in circulation.[2] The technological convergence has resulted in establishing a new and

[1] Statista, "Smartphone Penetration Worldwide 2014-2021," https://www.statista.com/statistics/203734/global- smartphone-penetration-per-capita-since-2005/.
Statista, "Global Smartphones Installed Base 2008-2017 | Statistic," https://www.statista.com/statistics/371889/smartphone-worldwide-installed-base/.

[2] Statista, "Smartphone Penetration United States 2017-2023 | Statistic," https://www.statista.com/statistics/201184/percentage-of-mobile-phone-users-who-use-a-smartphone-in-the-us/.
Statista, "Smartphones Installed Base in the US 2013-2022 | Statistic," https://www.statista.com/statistics/619838/smartphones-installed-base-in-the-us/.

prominent smartphone industry sector, worth over $350 billion globally, according to some estimates.[3]

Technological convergence may present a range of issues where Congress may take legislative and/or oversight actions. Three selected issue areas associated with technological convergence are regulatory jurisdiction, digital privacy, and data security. First, merging and integrating multiple technologies from distinct functional categories into one converged technology may pose challenges to defining regulatory policies, roles, and responsibilities. Determining oversight jurisdictions and regulatory authorities for converged technologies may become complicated as the boundaries that once separated single-function technologies are blended together. In other words, delineating which policy authorizes which government agency to apply which standards to regulate which industry is no longer simple and straightforward. How Congress chooses to oversee certain industries and government agencies may also become complicated due to converging technologies that blur and blend existing categorical boundaries.

Second, digital privacy concerns stem from converged technologies' collection and usage of personal and machine data.[4] Technological convergence facilitates increasing consumption and collection of data, which poses potential digital privacy concerns for consumers. Data collection and usage are tied to digital privacy issues because a piece or aggregation of information could identify an individual or reveal patterns in their activities. Converged technologies leverage large volumes of data to try to improve the user experience by generating more tailored and anticipatory results. This data can also potentially be used to identify,

---

[3] Statista, "Global Smartphone Revenue 2011-2018 | Statistic," https:// www.statista.com/ statistics/687476/global- smartphone-revenues/.
Arjun Kharpal, "6 Billion Smartphones Will Be in Circulation in 2020: Report," *CNBC*, January 17, 2017, https://www.cnbc.com/2017/01/17/6-billion-smartphones-will-be-in-circulation-in-2020-ihs-report.html.

[4] Personal data refers to information that pertains to a specific individual person. Examples of personal data include name, social security number, email address, phone number, home address, fingerprints, and genetics information. Machine data refers to information generated by a machine—such as a computer, application, sensors, or a device—based on operational activities. Examples of machine data include sensor readings, network data for communication protocols, and web logs.

locate, track, and monitor an individual without the person's knowledge. The same data can potentially be sold to third-party entities without an individual's awareness. As the use of converged technologies continues to propagate, digital privacy issues will likely remain central to the policy debate.

Third, data security concerns are often associated with smart devices.[5] As devices are able to interconnect, the convenient ubiquitous features may create vulnerabilities that could be exploited by malicious actors. Data security, a component of cybersecurity, protects data from unauthorized access and use. Along with digital privacy, data security is a pertinent issue for converged technologies, which generate and consume large volumes of data. Technological convergence poses three potential data security concerns: increased number of access points susceptible to cyberattacks, linkage to physical security, and theft of data.

The first section of this chapter describes technological convergence along with closely associated media convergence and network convergence. The report uses the Internet of Things (IoT) and smart home devices as primary examples.[6] Of these three convergences, consumers most often directly engage with converged technologies. In contrast, general consumers may not have the same level of engagement or understanding of media and network convergences, as they often occur in the background.

The second section of this chapter presents regulatory, digital privacy, and data security issues pertaining to technological convergence. The current state, challenges, and recent legislative activities are discussed.

---

[5] Many technological convergence devices are called "smart" devices, which are electronic devices that have sensors and are connected to a network and/or other devices for sharing information and interacting with users. Sophisticated smart devices have machine learning algorithms that offer the ability to learn from past data and user preferences.

[6] Media convergence refers to content that is made available through multiple forms, formats, and access points on multiple platforms. Network convergence refers to a single network infrastructure that handles and distributes multiple types of media. The IoT is a system of interrelated devices that transfer data over a network among connected devices without requiring human-to-human or human-to-computer interaction. These topics are further discussed in subsequent sections of this chapter.

The third section of this chapter concludes with potential considerations for Congress. An overarching consideration for regulatory, digital privacy, and data security issues may be determining the role, if any, of the federal government in an environment where technological evolution changes quickly and continues to disrupt existing frameworks. Policies governing these three issues—regulations, digital privacy, and data security—may be of interest to Congress as well as other stakeholders, including U.S. government agencies, commercial entities, and the general public.

Description of Technological Convergence

"Technological convergence" is a concept whereby merging, blending, integration, and transformation of independent technologies leads to a completely new converged technology. This broad, complex concept encompasses a wide range of technologies, including IoT and smart home devices. When a converged technology emerges, it often replaces single-function technologies or renders them obsolete. In this sense, technological convergence can be viewed as a progression or evolution of technology.[7]

A discussion of technological convergence in isolation is difficult because technological convergence is closely associated with media convergence and network convergence. Technological, media, and network convergences are interdependent, but each possesses subtle distinctions. These three terms are often used interchangeably, further complicating the discussion of an already complex topic. **Figure 1** illustrates relationships between technological, media, and network convergences.

---

[7] Dong Hee Shin, Won-Yong Kim, and Dong-Hoon Lee, "Convergence Technologies and the Layered Policy Model: Implication for Regulating Future Communications," in *Annual Meeting* (International Communication Association, United States: International Communication Association, 2006), http://web.b.ebscohost.com/ehost/detail/detail?vid=3&sid=00dd6275-99fe-4a5a-a32d-4ccb6b08620c%40pdc-v-essmgr03&bdata=JnNpdGU9ZWhvc3QtbGl2ZSZzY29wZT1zaXRl#AN=27203990&db=ufh.

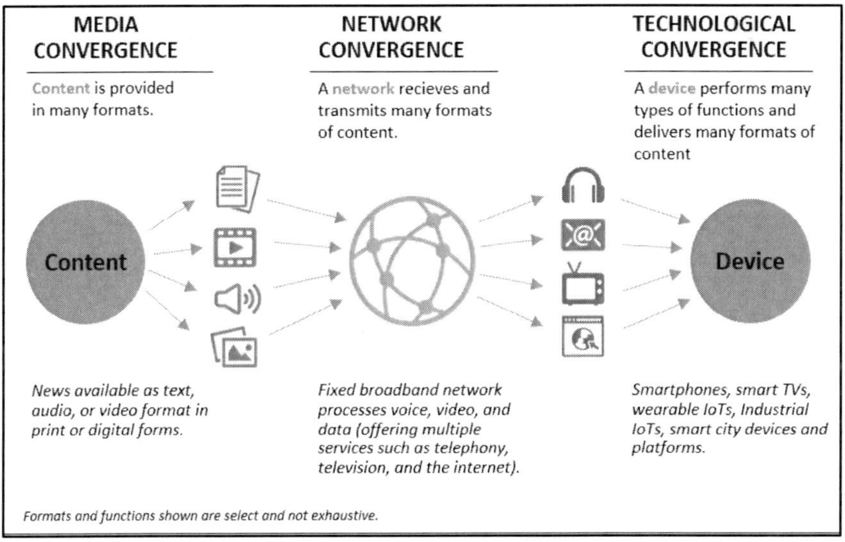

Source: CRS.

Figure 1. Technological, Media, and Network Convergences.

Technological Convergence

This occurs when the *functions* of different technologies are merged and interoperate as a single unit. A converged unit can typically process multiple types of media that correspond to each technology that merged. Technological convergence includes devices and systems that interface with end users. For example, a user interacts with converged devices, such as a smart television (TV), to access the contents that are distributed over a network. A smart TV has combined the functions of a traditional TV, a computer, and several other devices that used to have one specific purpose. In addition to displaying over-the-air broadcast TV channels, smart TVs interface with users to surf the internet, view photos taken from smartphones and stored in the "cloud," display feeds from home security cameras connected to a network, play music, notify users of incoming calls and messages, and allow video teleconferencing.[8] Smart TVs can process a variety of formats of media to perform multiple functions.

Media Convergence

This refers to *content* that is made available through multiple forms, formats, and access points. Media convergence proliferated as analog mediums of communication became digitized. For example, the contents on a newspaper used to be available only in print. The same content is currently available in both print and digital forms, as text, visual, and/or audio formats, and through multiple devices and platforms including social media.

Network Convergence

This refers to a single *network infrastructure* that handles and distributes multiple types of media. Network convergence became prominent when telecommunications and information networks integrated; it became prevalent when mobile cellular communications incorporated access to the internet and made it widespread. For example, today's cable companies process information in forms of voice, video, and data on a single network and often offer their services as a bundle package (e.g., phone, television, and internet services). Similarly, cellular networks, which distribute information to and from mobile devices and fixed platforms, process voice, video, and data.

Prior to network, media, and technological convergences, a separate, independent network was dedicated to handling and distributing one particular type of media that was processed by a single-function device. For example, a telephone network distributed audio information (i.e.,

[8] Cloud storage refers to storing digital data on remote servers and accessing the data through the internet. Because data are not stored in a single, fixed, isolated location, and can be accessed from anywhere, cloud storage gives the impression of storing data on a non-concrete, flexible, fluid, mobile "cloud."

voice) between telephone handsets. A broadcasting network delivered video to television sets. Convergence removes such pairing (i.e., "decouples") between media, network, and device.[9] Decoupling gives convergence its versatility, flexibility, and complexity.

Characteristics of Smart Devices

Many technological convergence devices are called "smart" devices, which often include IoT devices. (Examples of IoT devices are discussed in following sections.) Despite a wide range of applications, smart converged technologies share key characteristics

- Smart devices can execute multiple functions to serve blended purposes;
- Smart devices can collect and use data in various formats and employ machine learning algorithms[10] to deliver optimized and enhanced user experience; and
- Smart devices are connected to a network directly and/or are interconnected with other smart devices, offering ubiquitous access to users from anywhere on any platform.

These key characteristics may present potential policy questions for Congress, including the following:

---

[9] Corinna Peil and Sergio Sparviero, "Media Convergence Meets Deconvergence," in Media Convergence and *Deconvergence*, ed. Sergio Sparviero, Corinna Peil, and Gabriele Balbi, Global Transformations in Media and Communication Research—A Palgrave and IAMCR Series (Cham: Springer International Publishing, 2017), 3-30, https://doi.org/10.1007/978-3-319-51289-1_1.

[10] Machine learning algorithms examine historical information and user preferences, extract patterns, and attempt to predict or anticipate a user's need. The predictive capability improves with more historical data and iterations of adjusting rules or model parameters. According to the Merriam-Webster dictionary, an algorithm is "a procedure for solving a mathematical problem (as of finding the greatest common divisor) in a finite number of steps that frequently involves repetition of an operation." Many different mathematical principles form the basis of machine learning algorithms, executed as a software program that leverage computing power and executes functions without explicit commands.

- Who will provide oversight and how will regulatory authorities be applied to technologies that serve multiple functions or that do not belong to an established category?
- How should consumer data be collected and used to protect digital privacy without limiting technology innovation?
- How to shape data security practices to safeguard personal information and physical security from malicious actors?

An Overview of Internet of Things (IoT)

The IoT is a common example of technological convergence. The IoT is a system of devices that are connected to a network and each other, exchanging data without necessarily requiring human- to-human or human-to-computer interaction.[11] In other words, IoT is a collection of electronic devices that can share information among themselves (e.g., smart home devices). The IoT possess all three characteristics of converged technologies: multiple functions, data collection and use, and ubiquitous access. Various categories of IoT include industrial Internet of Things, Internet of Medical Things, smart city infrastructures, and smart home devices.

IoT industry is a growing market both globally and in the United States.[12] According to some estimates, in 2018, the IoT retail market in the United States was almost $4 billion,[13] and over 700 million consumer IoT

---

[11] Samuel Greengard, *The Internet of Things*, The MIT Press Essential Knowledge Series (Cambridge, Massachusetts: The Massachusetts Institute of Technology Press, 2015).
IoT Agenda, "What Is Internet of Things (IoT)?—Definition from WhatIs.Com," https:// internetofthingsagenda.techtarget.com/definition/Internet-of-Things-IoT.

[12] Statista, "IoT Market Size Worldwide 2016-2020 | Statistic," https:// www.statista.com/ statistics/764051/iot-market- size-worldwide/.
Daniel Alsen, Mark Patel, and Jason Shangkuan, "The Future of Connectivity: Enabling the Internet of Things," McKinsey and Company, https://www.mckinsey.com/featured-insights/ internet-of-things/our-insights/the-future-of-connectivity-enabling-the-internet-of-things.
Statista, "IoT Infrastructure Market by Application in US 2016-2024 | Statistic," https:// www.statista.com/statistics/761278/iot-infrastructure-market-by-application-in-us/.

[13] Statista, "IoT Hardware in US Retail Market 2014-2025 | Statistic," https://www.statista.com/ statistics/688756/iot-in- retail-market-in-the-us/.

devices were in use in 2017 in the United States.[14] **Figure 2** illustrates global revenue of the IoT from 2012 to 2018, according to Statista, a company that consolidates statistical data, based on information from IC Insights.[15] In 2018, consumer IoT devices, such as wearable and connected smart home devices, generated over $14 billion globally. The connected cities category, or smart cities, was the largest (41%) of 2018 global IoT revenue.

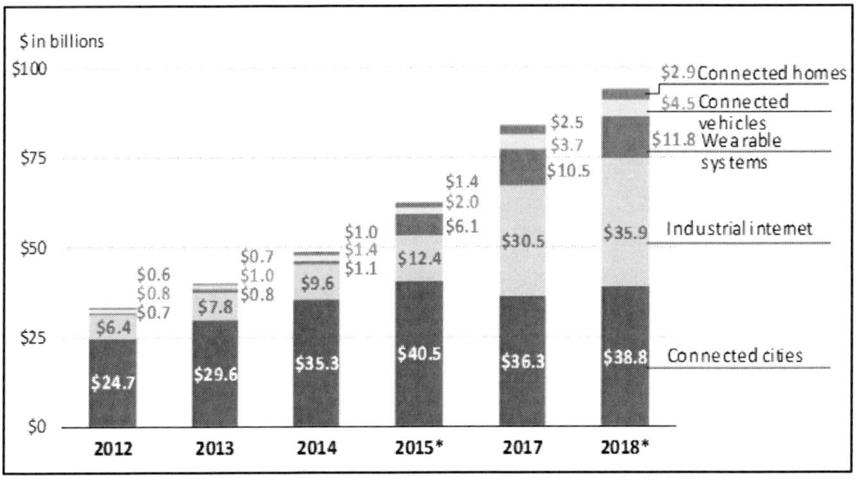

Source: CRS created based on data from Statista, "IoT Subsystems Revenue Worldwide 2012-2018 | Statistic," https://www.statista.com/statistics/503466/iot-subsystems-revenue-worldwide/.

Notes: The source for this figure did not have the information for 2016 and applied estimations for 2015 and 2018 (with *). The Connected cities category refers to smart cities; the Industrial internet category refers to industrial internet of things; the Wearable systems category includes consumer devices that register and provide personal health information; and the Connected homes category refers to smart homes.

Figure 2. Internet of Things Subsystems Revenue Worldwide from 2012 to 2018.

[14] Statista, "IoT Devices in Use by Category in US 2017 | Statistic," https://www.statista.com/statistics/757717/iot- consumer-product-installed-base-in-the-us-by-category/.

[15] Statista, "IoT Subsystems Revenue Worldwide 2012-2018 | Statistic," https://www.statista.com/statistics/503466/iot- subsystems-revenue-worldwide/.

The industrial Internet of Things, such as smart factories, had the biggest growth in terms of global revenue between 2017 and 2018 among the different categories of the IoT. An estimate of various IoT markets by McKinsey also shows the industrial IoT as potentially increasing the most by 2025 compared to other IoT systems.[16] The development, application, and usage of IoT will likely continue to grow with Fifth-Generation (5G) Technologies cellular service, which will allow a larger number of devices to be connected simultaneously to a network, supporting not only consumer but industrial use of IoT devices and systems.[17]

IoT devices are used in many different fields and serve a variety of functions. The IoT encompasses a broad range of applications. Selected categories of IoT devices are discussed below.

Industrial Internet of Things (IIoT)

Examples of commercial application of the IoT can be found in the manufacturing industry. Referred to as industrial Internet of Things (IIoT), networked machines in a production facility can communicate and share information to improve efficiency, productivity, and performance.[18] The application of IIoT can vary significantly, from detecting corrosion inside a refinery pipe to providing real-time production data.[19] Also, IIoTs can enable a variety of industries, such as manufacturing, chemicals, food and beverage, automotive, and steel, to transform their operations and potentially yield financial benefits.[20] Currently in North America, there are more consumer IoT connections than IIoT connections, but this may

---

[16] James Manyika et al., "Unlocking the Potential of the Internet of Things," McKinsey and Company, https://www.mckinsey.com/business-functions/digital-mckinsey/our-insights/the-internet-of-things-the-value-of-digitizing-the-physical-world.

[17] CRS Report R45485, *Fifth-Generation (5G) Telecommunications Technologies: Issues for Congress*, by Jill C. Gallagher and Michael E. DeVine.

[18] Lea Bolz, Heike Freund, Tarek Kasah, and Bodo Koerber, "IIoT Platforms for Industrial Equipment and Machinery Players," McKinsey and Company, https://www.mckinsey.com/industries/advanced-electronics/our-insights/iiot- platforms-the-technology-stack-as-value-driver-in-industrial-equipment-and-machinery.

[19] GE Digital, "Everything You Need to Know about IIoT," https://www.ge.com/digital/blog/everything-you-need- know-about-industrial-internet-things.

[20] Ibid.

change in the future.[21] Incorporation of IIoT and analytics is considered by some as the Fourth Industrial Revolution (4IR).[22]

Internet of Medical Things (IoMT)

Some experts project the use of Internet of Medical Things (IoMT) is increasing.[23] IoMT devices, such as heart monitors and pace makers, collect and send a patient's health statistics over various networks to health care providers for monitoring, remote configuration, and preventions. In 2017, over 300 million IoT devices in the medical sector were connected worldwide, and, in 2018, over 400 million devices were connected.[24] At a personal health level, wearable IoT devices, such as smart watches and fitness trackers, can track a user's physical activities, basic vitals, and sleeping patterns. In 2017, over 40 million fitness tracker IoT were in use in the United States.[25]

Smart Cities

IoT devices and systems in transportation, utilities, and infrastructure sectors may be grouped under the category of "smart city."[26] An example of utilities IoT in a smart city is "smart" grid and meters for electricity, water, and gas where sensors collect and share customer usage data to enable the central control system to optimize production and distribution to meet demand real-time.[27] An example of transportation IoT in a smart city is fare readers and status trackers or locaters that interface across all public

---

[21] Statista, "IoT Connections North America 2017-2025 | Statistic," https://www.statista.com/statistics/933099/iot- connections-north-america/.

[22] PricewaterhouseCoopers, "4IReady," https://www.pwc.com/us/en/library/4ir-ready.html.

[23] Bernard Marr, "Why the Internet of Medical Things (IoMT) Will Start to Transform Healthcare in 2018," *Forbes*, https://www.forbes.com/sites/bernardmarr/2018/01/25/why-the-internet-of-medical-things-iomt-will-start-to-transform-healthcare-in-2018/.

[24] Statista, "Global Connected IoT Devices by Sector 2017 and 2018 | Statistic," https://www.statista.com/statistics/748737/worldwide-connected-iot-devices-by-sector/.

[25] Statista, "IoT Devices in Use by Category in US 2017 | Statistic," https://www.statista.com/statistics/757717/iot- consumer-product-installed-base-in-the-us-by-category/.

[26] "Smart Cities and Artificial Intelligence," Brookings (blog), January 25, 2019, https://www.brookings.edu/events/smart-cities-and-artificial-intelligence/.

[27] Andrew Meola, "IoT for Utilities: Smart Water, Gas and Electric Utilities Coming Soon," *Business Insider*, https://www.businessinsider.com/internet-of-things-utilities-water-electric-gas-2016-10.

transportation platforms.[28] Columbus, OH's winning proposal for the Department of Transportation's (DOT) Smart City Challenge of 2016, included connected infrastructure that interacts with vehicles, trip planning and common payment system across multiple transit system, and electric autonomous vehicles and shuttles.[29] Other finalists of the DoT Smart City Challenge were Austin, TX; Denver, CO; Kansas City, MO; Pittsburgh, PA; Portland, OR; and San Francisco, CA.[30] Smart cities is currently the largest segment of IoT in terms of revenue.[31]

Smart Home

Consumer product IoT devices used in homes and buildings are often grouped under the "smart home" category. Included in this categories are smart appliances, smart TV, smart entertainment systems, smart thermostats, and network-connected light bulbs, outlets, door locks, door bells, and home security systems.[32] These smart home IoT devices are connected to a single network and can be controlled remotely over the internet. Eight of 11 categories of consumer IoT devices used in 2017 were related to smart home.[33] In 2018, the size of the global smart home market was estimated to be over $30 billion.[34]

---

[28] U.S. Department of Transportation, "Smart City Challenge," September 28, 2016, https://www.transportation.gov/smartcity.

[29] Smart Columbus, "Home | SmartColumbus," https://smart.columbus.gov; U.S. Department of Transportation," The Winner: Columbus, Ohio," September 28, 2016, https://cms.dot.gov/smartcity/winner.

[30] U.S. Department of Transportation, "Round Two: Seven Finalists Create Plans To Implement Their Visions," September 28, 2016, https://cms.dot.gov/smartcity/7-finalists-cities.

[31] Statista, "IoT Subsystems Revenue Worldwide 2012-2018 | Statistic," https://www.statista.com/statistics/503466/iot- subsystems-revenue-worldwide/.

[32] Statista, "Smart Home—Worldwide | Statista Market Forecast," https://www.statista.com/outlook/279/100/smart- home/worldwide.

[33] Statista, "IoT Devices in Use by Category in US 2017 | Statistic," https://www.statista.com/statistics/757717/iot- consumer-product-installed-base-in-the-us-by-category/.

[34] Statista, "Global Smart Home Market Size 2016-2022 | Statistic," https://www.statista.com/statistics/682204/global- smart-home-market-size/.
Market size is different from revenue. Market size is the potential revenue—an estimated, possible, total income that can be generated from all potential customers, which compose a segment of a market. Revenue is the total positive cash flow from sales of goods or services in a given timeframe.

An Example: Smart Home

A smart home contains a collection of consumer IoT devices intended for personal use where user experience is improved by connecting various features of a house to a network. For example, smart home IoT devices may be interconnected to each other and to a central control system for a home with voice interface, often referred to as a virtual assistant. Commonly known examples of virtual assistants are Amazon's Alexa, Apple's Siri, Google Assistant, Microsoft's Cortana, and Samsung's Bixby. A virtual assistant is a platform that can manage and relay information to smart home devices based on user-established criteria.

Moreover, a smart home may have a doorbell with a video camera and a speaker that allows a user to see who is at the door and to speak to the person at the door from anywhere over the internet. A smart home may have a smart door lock that can be locked and unlocked remotely. In addition, the thermostat, lights, electrical outlets, and appliances in a smart home may be remotely controlled by a user over the internet. A smart appliance, such as a smart refrigerator that is networked, can use its sensors to identify items and can notify a user based on set criteria, such as restocking alerts or suggested recipes.

Some smart home devices resemble traditional devices, but with cross-over functions or networking abilities. Examples include smart lightbulbs, smart electrical outlet plugs, smart TV, and smart appliances. Some smart home devices are establishing a new category of industry segment that did not exist previously. An example is Amazon's Echo products with virtual assistant Alexa as voice user interface. Whether it is the former (evolutionary technologies) or the latter (new/revolutionary technologies), the smart home industry is fast emerging and growing.[35]

Smart home devices, which are a type of IoT, possess the three characteristics of converged technologies: multiple or blended functions, collection and use of data, and ubiquitous access through network connection. Thus, potential policy interests associated with technological

---

[35] Statista, "Smart Home Report 2019," https://www.statista.com/study/42112/smart-home-report/.

convergence can be also observed in smart home devices. Potential smart home issues for Congress include the following.

Congress may decide it is necessary to resolve oversight jurisdictions and regulatory authorities of smart home devices, especially for products like virtual assistants, which may not belong to an established category of technology. The mission of the Federal Trade Commission (FTC) includes both protecting consumers and promoting business competition.[36] Congress may choose to review the FTC's current authorities to ensure that they are sufficient to oversee emerging smart home technologies. In addition, potentially deconflicting or harmonizing jurisdictions may be discussed if other federal government organizations and their mission are impacted by emerging smart home technologies.

Congress may decide that new or expanded policies are necessary to protect consumer digital privacy, including personal data that are collected and used by smart home devices, such as a smart TV, in private spaces, such as a user's home. Although the FTC does promote a level of digital privacy through its consumer protection authorities, emerging digital privacy issues are linked to practices that are legal as opposed to fraud, theft, or other malicious activities.[37] Congress may examine whether a federal law that comprehensively addresses personal digital privacy is necessary or an expansion of the FTC's consumer data protection authorities is required.

Emerging smart home technologies may further necessitate safeguarding data from malicious actors. In addition to collecting and using personal data, smart home devices bridge physical security and cybersecurity. Malicious actors may have more means to exploit a user's information and home through smart home devices, which offer ubiquitous access as a key convenience feature. Whether current policies adequately

---

[36] Federal Trade Commission, "What We Do," June 7, 2013, https://www.ftc.gov/about-ftc/what-we-do. Federal Trade Commission, "A Brief Overview of the Federal Trade Commission's Investigative and Law Enforcement Authority," June 7, 2013, https://www.ftc.gov/about-ftc/what-we-do/enforcement-authority.

[37] Federal Trade Commission, "Data and Visualizations," October 15, 2018, https://www.ftc.gov/enforcement/data- visualizations.

addresses data, cyber, and physical security concerns may also be considered.

SELECTED ISSUES ASSOCIATED WITH TECHNOLOGICAL CONVERGENCE

Regulation, digital privacy, and data security are three selected issues associated with technological convergence that may be of interest to many stakeholders, including Congress. As identified in the smart home example in the previous section, each of these three issues is discussed further in subsequent subsections. The three selected issues are tied to the three characteristics of converged technologies discussed previously in the "Characteristics of Smart Devices" section. First, convergence of technologies blend and blur existing categorical distinctions for each technology because a converged technology can perform multiple functions. Second, technological convergence consumes, collects, and generates a large volume of both personal and machine data. Third, converged technologies allow ubiquitous access points to the end users.[38] These characteristics are typically observed as a result of decoupling the devices from media and network.

Regulatory Issues

Congress may consider policies that address blending standards and boundaries as converged technologies and companies merge and replace traditionally independent and distinct categories. Policy issues may include oversight jurisdictions, regulatory authorities, and commercial competitiveness since a converged technology could fall within multiple domains. An example may be delineating the Federal Communications and Commission's (FCC) and the FTC's authorities on convergence

[38] Ubiquitous access refers to the capability of remotely accessing a device of a system anytime, from anywhere on any platform.

technologies as more devices and services become mobile and wirelessly connected.[39]

Merging and integrating multiple technologies from distinct functional categories into one converged technology pose challenges to regulatory policies and responsibilities. Determining oversight jurisdictions and regulatory authorities for converged technologies becomes unclear as the boundaries that once separated single-function technologies blend and blur together. A challenge for policymakers may be in delineating which government agency and which policies and standards would best apply to certain technologies or certain industries. Where there were once clear lines of authority by industry or media type (e.g., voice, video, data), they are no longer simple and straightforward for technologies where these functionalities have converged. How Congress oversees which industries and government agencies may become complicated due to converging technologies that blend existing categorical boundaries. Congress may decide that it is necessary for specific legislative committees to effectively oversee a converged technology that serves multiple functions. As a result, the alignment of converged technologies to regulatory authorities may shift as technologies evolve.

The complexities in setting regulatory jurisdiction can be further subdivided into regulating converging *technologies* and regulating evolving technology *companies*. They are discussed below.

Regulating Converging Technologies

Regulating a converging technology, which is a result of blending or integrating multiple technologies, can be challenging. This is because (1) the one-to-one relationship between a converging technology and a regulatory entity is no longer clear, and (2) a converging technology may create a new sector where a regulatory entity has not been identified.

Initially, the standards and oversight policies for a specific technology were established independently. They were not necessarily developed with merging or interoperability in mind. For example, telephony (when

[39] CRS In Focus IF10955, *Access to Broadband Networks: Net Neutrality*, by Angele A. Gilroy.

providing voice), cable TV (when providing video), and mobile cellular technologies each follow their respective standards, and these services were regulated by policies specific to each type. When a converged technology utilizes differing communications technologies, it may be required to adhere to multiple standards and regulations.

In such cases, multiple agencies may need to regulate a single converged technology. This may require extended timelines for regulatory reviews. Industry may incur additional costs to meet standards and reporting requirements for converged technologies.

In other situations, as technologies converge, the outcome may yield a completely new technology for which a regulatory category did not previously exist. Examples include social media, IoTs, and virtual assistants. Without a clear regulatory and oversight framework in place, new converged technologies may be left unregulated, partially regulated, or regulated under a newly developed framework. They could also be left to self-regulate by the industry; or they could be overlooked as governing bodies remain indeterminate on which jurisdictional boundaries need to be stretched to cover emerging technology fields.

Regulating Evolving Companies

Regulating companies that offer converged technologies is challenging because the services and product lines evolve and expand such that they do not fall within a single category. Although diversification is considered normal business practice, technological convergence broadens the operational range for companies, spanning multiple industry sectors. Antitrust concerns could arise, or companies may not be subjected to the same level of oversight and regulation due to lack of classification.

For example, companies such as Amazon, Apple, and Google each offer smart home devices and platforms. Some of these devices, such as a smart doorbell with a video camera, smart doors and locks, and networked contact sensors and video cameras, may function as home security devices. Many of these products are bundled as a starting kit for home security. However, these technology convergence companies may not be required to follow state and local regulations as traditional home security companies that provide monitored security service do.[40]

Another example discussed widely in Congress is social media—whether social media companies should be classified as information technology companies, as advertising and marketing firms, as communications platforms, or as the press. As converged technologies and associated companies straddle or fall between jurisdictional boundaries, regulatory roles and responsibilities become more complex.

Digital Privacy Issues

Congress may be interested in digital privacy concerns of converged technologies, which often collect and use personal information and machine data as they directly interface with end-users. Current federal laws protect certain types of data pertaining to privacy by specifying collection, storage, use, and dissemination practices. As converged technologies generate and innovatively leverage more types and volumes of data that can identify, locate, or track a person, consumer concerns for protecting digital privacy may intensify.

Technological convergence facilitates increased consumption and collection of data, posing potential digital privacy concerns for consumers. Data collection and usage are tied to digital privacy issues because a piece or aggregation of information could identify an individual or reveal

[40] Monitored security systems are actively observed by a professional home security company who take predetermined actions based on alerts and activities. Unmonitored or self-monitored security systems are managed by the user. Furthermore, professionally installed home security systems are set up by licensed technicians using equipment offered by a professional home security company. Do-it-yourself home security systems are installed by the user by purchasing a kit or sensor devices.
"What Organization Regulates Home Security Companies?," *SafeWise* (blog), https://www.safewise.com/home- security-faq/who-regulates-security companies/; "What Is a Security System and How Does It Work?," *SafeWise* (blog), https://www.safewise.com/home-security-faq/how-do-security-systems-work/; "The Difference Between Monitored and Unmonitored Security Systems," *SafeWise* (blog), https://www.safewise.com/home-security-faq/ monitored-unmonitored-systems/; and "Is a DIY Security System Better Than a Professionally Installed One?," *SafeWise* (blog), https://www.safewise.com/home-security-faq/diy-vs-professional/.

patterns in their activities. Converged technologies leverage large volumes of data to try to improve the user experience by generating more tailored and anticipatory results. However, such data can potentially identify, locate, track, and monitor an individual without the person's knowledge. As the use of converged technologies continues to propagate, digital privacy issues will likely remain central.

Current Data Protection Laws

While a federal law that comprehensively addresses digital privacy does not currently exist, many laws are in place to protect certain types of data and their impact on specific aspects of privacy. Current U.S. data protection laws include the following.

> Gramm-Leach-Bliley Act (GLBA): The GLBA imposes several data protection obligations on financial institutions. These obligations are centered on a category of data called "consumer" "nonpublic personal information" (NPI), and generally relate to: (1) sharing NPI with third parties, (2) providing privacy notices to consumers, and (3) security NPI from unauthorized access.
>
> Health Insurance Portability and Accountability Act (HIPAA): Under the HIPAA, the Department of Health and Human Services (HHS) has enacted regulations protecting a category of medical information called "protected health information" (PHI). These regulations apply to health care providers, health plans, and health care clearinghouses (covered entities), as well as certain "business associates" of such entities. The HIPAA regulations generally speak to covered entities': (1) using or sharing of PHI, (2) disclosure of information to consumers, (3) safeguards for securing PHI, and (4) notification of consumers following a breach of PHI.
>
> Fair Credit Reporting Act (FCRA): The FCRA covers the collection and use of information bearing on a consumer's creditworthiness. FCRA and its implementing regulations govern the activities of three categories of entities: (1) credit reporting agencies (CRAs), (2) entities furnishing information to CRAs (furnishers), and (3) individuals who use credit reports issued by CRAs (users). In contrast to HIPAA or GLBA, there are no privacy provisions in FCRA requiring entities to provide notice to a consumer or to obtain his opt-in or opt-out consent before collecting or disclosing the consumer's data to third parties. FCRA further has no data security provisions requiring entities to maintain safeguards to protect consumer information from unauthorized access. Rather, FCRA's requirements generally focus on ensuring that the consumer information

reported by CRAs and furnishers is accurate and that it is used only for certain permissible purposes.

The Communications Act: The Communications Act of 1934 (Communications Act or Act), as amended, established the Federal Communications Commission (FCC) and provides a "comprehensive scheme" for the regulations of interstate communication. [T]he Communications Act includes data protection provisions applicable to common carriers, cable operators, and satellite carriers.

Video Privacy Protection Act (VPPA): The VPPA was enacted in 1988 in order to "preserve personal privacy with respect to the rental, purchase, or delivery of video tapes or similar audio visual materials." The VPPA does not have any data security provisions requiring entities to maintain safeguards to protect consumer information from unauthorized access. However, it does have privacy provisions restricting when covered entities can share certain consumer information. Specifically, the VPPA prohibits "video tape service providers"—a term that includes both digital video streaming services and brick-and-mortar video rental stores—from knowingly disclosing [personally identifiable information] (PII) concerning any "consumer" without that consumer's opt-in consent. The VPPA does not empower any federal agency to enforce violations or the Act and there are no criminal penalties for violations, but it does provide for a private right of action for persons aggrieved by the Act.

Family Educational Rights and Privacy Act (FERPA): The FERPA creates privacy protections for student education records. "Education records" are defined broadly to generally include any "materials which contain information directly related to a student" and are "maintained by an educational agency or institution." FERPA defines an "educational agency of institution" to include "any public or private agency or institution which is the recipient of funds under any applicable program." FERPA generally requires that any "educational agency or institution" (i.e., covered entities) give parents or, depending on their age, the student (1) control over the disclosure of the student's educational records, (2) an opportunity to review those records, and (3) an opportunity to challenge them as inaccurate.

Federal Securities Laws: While federal securities statutes and regulations do not explicitly address data protection, two requirements under these laws have implications for how companies prevent and respond to data breaches. First, federal securities laws may require companies to adopt controls designed to protect against data breaches. Second, federal securities laws may require companies to discuss data breaches when making required disclosures under securities laws.

Children's Online Privacy Protection Act (COPPA): The COPPA and the FTC's implementing regulations regulate the online collection and use of children's information. Specifically, COPPA's requirements apply to: (1) any "operator" of a website or online service that is "directed to children," or (2) any operator that has any "actual knowledge that it is collecting personal information from a child" (i.e., covered operators). Covered operators must comply with various requirements regarding data collection and use, privacy policy notifications, and data security.

Electronic Communications Privacy Act (ECPA): The ECPA was enacted in 1986, and is composed of three acts: the Wiretap Act, the Stored Communications Act (SCA), and the Pen Register Act. Much of ECPA is directed at law enforcement, providing "Fourth Amendment like privacy protections" to electronic communications. However, "ECPA's three acts also contain privacy obligations relevant to non-governmental actors. ECPA is perhaps the most compressive federal law on electronic privacy, as it is not sector-specific, and many of its provisions apply to a wide range of private and public actors. Nevertheless, its impact on online privacy has been limited. As some commentators have observed, ECPA "was designed to regulate wiretapping and electronic snooping rather than commercial data gathering," and litigants attempting to apply ECPA to online data collection have generally been unsuccessful.

Computer Fraud and Abuse Act (CFAA): The CFAA was originally intended as a computer hacking statute and is centrally concerned with prohibiting unauthorized intrusions into computers, rather than addressing other data protection issues such as the collection or use of data. Specifically, the CFAA imposes liability when a person "intentionally accesses a computer without authorization or exceeds authorized access, and thereby obtains… information from any protected computer." A "protected computer" is broadly defined as any computer used in or affecting interstate commerce or communications, functionally allowing the statute to apply to any computer that is connected to the internet.

Federal Trade Commission Act (FTC Act): The FTC Act has emerged as a critical law relevant to data privacy and security. As some commentators have noted, the FTC has used its authority under the Act to become the "go-to agency for privacy," effectively filling in gaps left by the aforementioned federal statutes. While the FTC Act was originally enacted in 1914 to strengthen competition law, the 1938 Wheeler-Lea amendment revised Section 5 of the Act to prohibit a broad range of unscrupulous or misleading practices harmful to consumers. The Act gives the FTC jurisdiction over most individuals and entities, although there are several exemptions. For instance, the FTC Act exempts common carriers, nonprofits, and financial institutions such as banks, savings and loan institutions, and federal credit unions.

Consumer Financial Protection Act (CFPA): Similar to the FTC Act, the CFPA prohibits covered entities from engaging in certain unfair, deceptive, or abusive acts. Enacted in 2010 as Title X of the Dodd-Frank Wall Street Reform and Consumer Protection Act, the CFPA created the

> Consumer Financial Protection Bureau (CFPB) as an independent agency within the Federal Reserve System. The Act gives the CFPB certain "organic" authorities, including the authority to take any action to prevent any "covered person" from "committing or engaging in an unfair, deceptive, or abusive act or practice" (UDAAP) in connection with offering or providing a "consumer financial product or service."

State laws, such as California Consumer Privacy Act (CCPA), and international laws, such as European Union's General Data Protection Regulations (GDPR), aim to provide a comprehensive guidance on digital privacy.[41]

The FTC Act and the Clayton Act are the primary statutes that give the FTC investigative, law enforcement, and litigating authority to protect consumers and promote competition (i.e., antitrust).[42] The FTC "has enforcement or administrative responsibilities under more than 70 laws."[43] The FTC's consumer protection mission currently focuses more on data security issues— such as identity theft, violation of Do Not Call or Do Not Track, and deceptive advertising—than digital privacy concerns associated with lawful activities.[44] While consumer protection and digital privacy are

---

[41] Additional information on foreign government policies, such as EU's GDPR and China's Security First can be found in CRS Report R45584, *Data Flows, Online Privacy, and Trade Policy*, by Rachel F. Fefer.

[42] Federal Trade Commission, "A Brief Overview of the Federal Trade Commission's Investigative and Law Enforcement Authority," June 7, 2013, https://www.ftc.gov/about-ftc/what-we-do/enforcement-authority.
The Clayton Act (15 U.S.C. 12 et seq.), which was enacted on October 15, 2914, comprises the U.S. antitrust law along with the FTC Act (15 U.S.C. 41 et seq.) and the Sherman Act (15 U.S.C. 1 et seq.). According to the FTC, "the Commission is charged under Sections 3, 7 and 8 of this Act with preventing and eliminating unlawful tying contracts, corporate mergers and acquisitions, and interlocking directorates. This Act was amended by the Robinson-Patman Act, Pub. L. No. 74-692, 49 Stat. 1526, codified at 15 U.S.C. §§ 13, 13b, and 21a, under which the Commission is authorized to prevent certain practices involving discriminatory pricing and product promotion. The Hart-Scott-Rodino Act (HSR), adding Section 7A of the Clayton Act, is listed separately." This information is available at https://www.ftc.gov/enforcement/statutes/clayton-act and at http://uscode.house.gov/view.xhtml?req=granuleid%3AUSC-prelim-title15-section12&edition=prelim.

[43] Federal Trade Commission, "Statutes Enforced or Administered by the Commission," https://www.ftc.gov/enforcement/statutes; "What We Do," June 7, 2013, https://www.ftc.gov/about-ftc/what-we-do.

[44] Additional information on Do Not Call or Do Not Track can be found in CRS Report R45070, *Protecting Consumers and Businesses from Fraudulent Robocalls*, by Patricia Moloney Figliola.

increasingly becoming synonymous, consumer protection law alone may not provide sufficient jurisdiction and authority to encompass digital privacy and data security issues for all data on all devices.

Data Privacy and Data Security

Digital privacy discussions often involve two closely associated topics: data privacy and data security. Data privacy is the governing of data collection, use, and sharing. Data security is protection of data from unauthorized or malicious actors. These two topics often differ in the lawfulness of activities, the intended use of data, and the effect on an individual.

Data security is an aspect of cybersecurity more so than privacy. Data security defends against illicit activities such as theft of data. Data security practices include proactive measures against cyberattacks and responsive measures such as sending notifications to affected individuals upon a data breach.

Data security issues typically involve actors whose intents are malicious, who carry out unlawful activities, and use data in ways that harm an individual. Examples include breaking into a database or sending spear phishing emails to steal identity and financial information.[45] Stolen identity and financial information are often exploited, causing financial damage to individuals and businesses. Privacy implications arise when personal information is compromised during a data security incident.

Data privacy practices determine how and to what extent data are collected, used, and with whom the data are shared. Data privacy sets the

---

Federal Trade Commission, "Data and Visualizations," October 15, 2018, https://www.ftc.gov/enforcement/data- visualizations.

[45] Phishing refers to general attempts by malicious actors to deceive victims to steal sensitive personal information such as credit card or financial information, passwords, and account credentials. Spear phishing is a targeted attempt aimed to steal sensitive information from a specific individual. In spear phishing, malicious actors often impersonate a trustworthy entity of a targeted victim.

scope for control of personal information—this may include data ownership and responsibilities of involved entities.[46]

Data privacy issues typically arise from lawful activities, but personal information may have been collected, used, or shared beyond given permission or awareness of an individual. The process or results may reveal aspects of an individual that were unexpected. Examples of data privacy issues include mobile apps and websites collecting and using an individual's online activity and location data to suggest targeted ads. In general, such activities are a lawful commercial marketing strategy, from which the customers may benefit in forms of enhanced user experience and discounts. But, these activities become an issue when they lack transparency (i.e., when customers are not aware of what information is collected on them, who shares the information with whom, and how the information is used and for what purpose). Individuals may experience that their rights to privacy have been violated when aggregation of information reveals highly targeted information that an individual did not anticipate.

Key aspects of data privacy—such as data collection, storage, sharing, access, and use—are not defined for digital data that are often leveraged by convergent technologies. These key aspects are defined only for certain types of information, such as medical and financial, where federal laws are in place. Similar guidance is limited or not available for other personal data, such as the following:

- Geolocation data[47]collected by apps;
- Contact information and other user-generated content on social media;
- Video recordings made by smart home IoT devices;
- Voice recordings made by virtual assistants; and

[46] Y.H. Wong and Humphry Hung, "Information Transparency and Digital Privacy Protection: Are They Mutually Exclusive in the Provision of E-services?," *Journal of Services Marketing* 23, no. 3 (May 22, 2009), pp. 154-164, https://doi.org/10.1108/08876040910955161.

[47] An example of geolocation data is Global Positioning System (GPS) coordinates, where the information can identify the physical location an object including an electronic device.

- Vitals and health data collected by fitness tracking wearable IoT devices.

Committees in both the House and the Senate of the 115th Congress held several hearings where technology companies were present as witnesses. Over a dozen bills were introduced in the 115th Congress to address various aspects of data privacy and security; but, none became a law.

Committees in both the House and the Senate of the 116th Congress have already held multiple hearings on privacy. Several bills were introduced by the 116th Congress to address data privacy concerns as they relate to technological convergence. These bills include the following:

- H.R. 1282 (Representative Bobby Rush), introduced on February 14, 2019, as the Data Accountability and Trust Act, would "require certain entities who collect and maintain personal information of individuals to secure such information and to provide notice to such individuals in the case of a breach of security involving such information…." This bill would define the term personal information; outline special requirements for information brokers; and assign specific responsibilities to the FTC to regulate commercial entities' data security policies and procedures for using and protecting personal information.
- S. 142 (Senator Marco Rubio), introduced on January 16, 2019, as the American Data Dissemination (ADD) Act of 2019, would "impose privacy requirements on providers of internet services similar to the requirement imposed on Federal agencies under the Privacy Act of 1974." This bill would require the FTC to submit recommendations for privacy requirements for internet service providers.
- S. 189 (Senator Amy Klobuchar), introduced on January 17, 2019, as the Social Media Privacy Protection and Consumer Rights Act of 2019, would "protect the privacy of users of social media and other online platforms." This bill would require commercial

entities with an online platform to clearly disclose their practices for personal data collection and use prior to obtaining user consents. This bill also outlines enforcement of privacy requirements by the FTC and the attorney general of each state.

- S. 583 (Senator Catherine Cortez Masto), introduced on February 27, 2019, as the Digital Accountability and Transparency to Advance (DATA) Privacy Act, would provide "digital accountability and transparency." This bill would require commercial entities to clearly disclose its privacy practices for various collected data. This bill also would require the FTC to enforce privacy practices to ensure that the minimum requirements are satisfied.

Data Brokers

According to the FTC, data brokers are

> companies that collect consumers' personal information and resell or share that information with others. Data brokers collect personal information about consumers from a wide range of sources and provide it for a variety of purposes, including verifying an individual's identity, marking products, and detecting fraud. Because these companies generally never interact with consumers, consumers are often unaware of their existence, much less the variety of practices in which they engage.[48]

The FTC classifies data brokers into three categories:[49]

1. Entities subject to the FCRA;
2. Entities that maintain data for marketing purposes; and

---

[48] Federal Trade Commission, "Data Brokers: A Call for Transparency and Accountability," May 2014, http://www.ftc.gov/system/files/documents/reports/data-brokers-call-transparency-accountability-report-federal-trade-commission-may-2014/140527databrokerreport.pdf.

[49] Federal Trade Commission, "Protecting Consumer Privacy in an Era of Rapid Change: Recommendations For Businesses and Policymakers," March 1, 2012, https://www.ftc.gov/reports/protecting-consumer-privacy-era-rapid-change-recommendations-businesses-policymakers.

3. Non-FCRA covered entities that maintain data for non-marketing purposes that fall outside of the FCRA.

The FCRA governs the activities of credit reporting agencies, such as Equifax, Experian, and TransUnion; entities furnishing information to credit reporting agencies; and individuals who use credit reports issued by credit reporting agencies.[50] These entities subjected to the FCRA fall within the first of the three categories of data brokers listed above. However, the FCRA does not have privacy or data security provisions.

Regarding the second and third categories of data brokers, the FTC report notes that "while the FCRA addresses a number of critical transparency issues associated with companies that sell data for credit, employment, and insurance purposes, data brokers within the other two categories remain opaque."[51] Data brokerage companies include Acxiom, Cambridge Analytica, Corelogic, Datalogix, Epsilon, Exactis, ID Analytics, Intelius, PeekYou, Rapleaf, and Recorded Future in addition to the "big three" credit reporting agencies (Equifax, Experian, and TransUnion).[52]

Many data brokers, which are conducting lawful activities, are self-regulated. As depicted in **Figure 3**, data brokerage companies purchase and aggregate information from various sources, which are also self-regulated. These sources include app developers, websites, and social media.[53]

---

[50] CRS Report R45631, *Data Protection Law: An Overview*, by Stephen P. Mulligan, Wilson C. Freeman, and Chris D. Linebaugh.

[51] Federal Trade Commission, "Data Brokers: A Call for Transparency and Accountability," May 2014, http://www.ftc.gov/system/files/documents/reports/data-brokers-call-transparency-accountability-report-federal-trade-commission-may-2014/140527databrokerreport.pdf.

[52] Privacy Rights Clearinghouse, "Data Brokers," https://www.privacyrights.org/data-brokers.

[53] Jennifer Valentino-DeVries et al., "Your Apps Know Where You Were Last Night, and They're Not Keeping It Secret," *New York Times*, December 10, 2018, sec. Business, https:// www.nytimes.com/interactive/2018/12/10/ business/location-data-privacy-apps.html.

Joseph Cox and Jason Koebler, "I Gave a Bounty Hunter $300. Then He Located Our Phone," *Motherboard* (blog), January 8, 2019, https://motherboard.vice.com/en_us/article/nepxbz/i-gave-a-bounty-hunter-300-dollars-located-phone-microbilt-zumigo-tmobile.

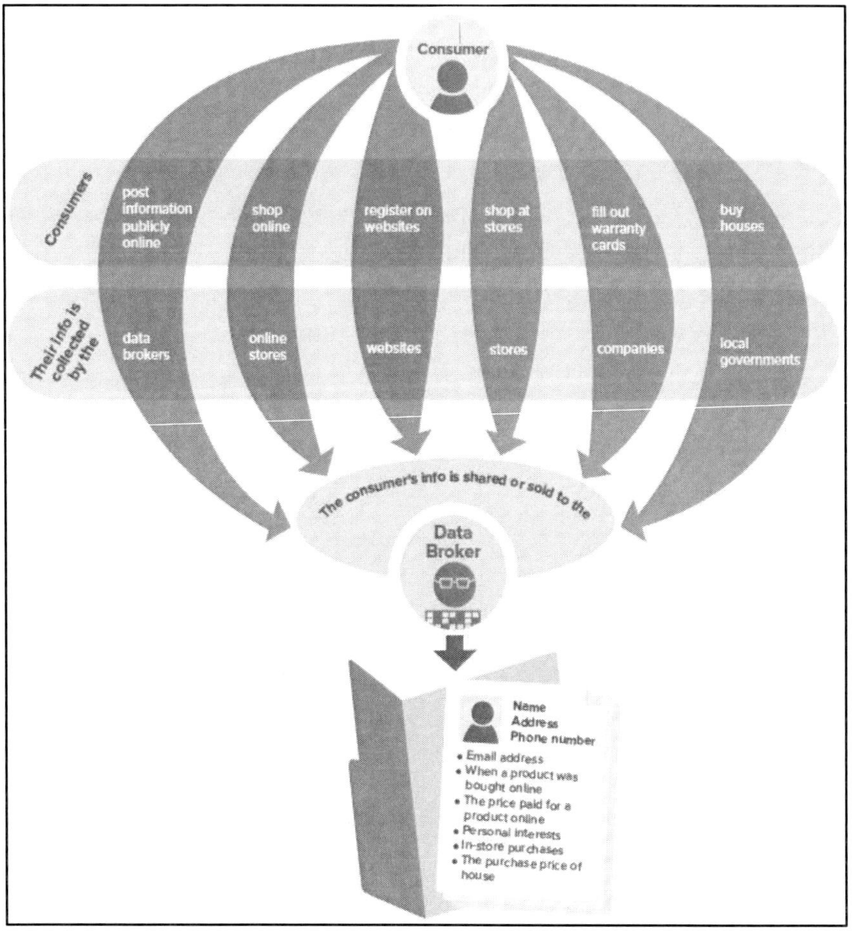

Source: Federal Trade Commission, "Data Brokers: A Call for Transparency and Accountability," May 2014, http://www.ftc.gov/system/files/documents/ reports/ data-brokers-call-transparency-accountability-report-federal- trade-commission-may-2014/140527databrokerreport.pdf.

Figure 3. Data Collection Online and Offline. As consumers go about their business, data brokers may collect information about them.

As technological convergence continues to proliferate, more data will likely be generated and consumed. Aggregations of seemingly simple and benign pieces of data when examined together could expose highly personal aspects in detail. Data brokers and entities that collect data could significantly impact digital privacy especially if individuals remain unaware of activities pertaining to their personal data.

Data Security Issues

Congress may be interested in data and physical security aspects of converged technologies because ubiquitous access equates to more possible entry points for both authorized and unauthorized users. This is often referred to as increase in attack surface. As more converged devices become connected to each other and to the internet, the overall impact of a compromise increases, along with the possibility of a cascading effect of a cyberattack. In policies, the requirements and responsibilities of data protection may be addressed separately from privacy concerns associated with legal use of personal data.

Data security, a component of cybersecurity, protects data from unauthorized access and use. Along with digital privacy, data security is a pertinent issue to technological convergence, which generates and consumes large volumes of data. Technological convergence poses a number of different types of potential data security concerns, including the following: potentially increased number of access points susceptible to cyberattacks, linkage to physical security, and theft of data.

Increased connectivity generally translates to increased risk of cyberattack.[54] Converged technologies, such as IoT devices, offer the users ubiquitous access: access from anywhere, at any time, using any device. While this is an extremely convenient characteristic, it also poses cybersecurity concerns. Multiple access points equate to increased points or opportunities for potential exploitation by malicious actors. This is often described as increased attack vectors, or broadening attack surface, which

[54] Dan Timpson, "The Internet of Things Is Becoming the Internet of Threats," Forbes, https://www.forbes.com/sites/forbestechcouncil/2018/10/24/the-internet-of-things-is-becoming-the-internet-of-threats/.

is a sum of attack vectors.[55] The same entry points a user may use for remote access can be exploited by an adversary to steal personal information. From the data security perspective, this is a tradeoff to consider between convenience and vulnerability.

Cybersecurity and physical security are directly linked through converged technologies. For example, when smart doors and smart locks are remotely controlled by a malicious actor through cyberattack, the physical security of that building also becomes compromised. The damage may not be limited to loss of digital content or information. Loss of personal data stored in the compromised location as well as personal security could be in jeopardy.

Potential loss or theft of personal data may be a data security concern for converged technologies because IoT devices often do not employ strong encryption at the device or user interface level. Not implementing strong encryption may be intentional due to associated benefits—it usually keeps the cost low, increases battery life of devices, minimizes memory requirements, reduces device size, and is easier to use or implement. This means, not only is the attack vector increased, but a system is also easier to break into. IoT devices may be the most vulnerable points of a system targeted by malicious actors for exploitation. Some experts note that IoT security currently lacks critical elements such as end-to-end security solutions, common security standards across the IoT industry, and customers' willingness to pay additional cost for enhanced security.[56]

Congressional Considerations for Technological Convergence

With relatively few policies in place for specifically overseeing technological convergence, Congress may consider potential policy options

---

[55] Pratyusa K. Manadhata, "An Attack Surface Metric" (School of Computer Science, Carnegie Mellon University, 2008), http://reports-archive.adm.cs.cmu.edu/anon/2008/CMU-CS-08-152.pdf.

[56] Harald Bauer, Ondrej Burkacky, and Christian Knochenhauer, "Security in the Internet of Things," McKinsey and Company, https://www.mckinsey.com/industries/semiconductors/our-insights/security-in-the-internet-of-things.

to address the issues discussed in this chapter. The fundamental policy considerations to identifying options may be determining the role, if any, of the federal government in overseeing technological convergence, digital privacy, and data security.

Regulatory Considerations

Regulating technological convergence may entail policies for jurisdictional deconfliction, harmonization, and expansion to address blended or new categories of technology. Currently, aspects of converged technologies may be regulated by different agencies based on the individual technologies that compose the convergence, but not as a whole. Regulating a converged technology as a whole can also be challenging because the combinations of technologies may generate too many possible outcomes. When converged technologies establish a new domain and fall outside of existing regulatory jurisdictions, they are often left to self-regulate.

Congress and the Administration could take a number of approaches in regulating technological convergence. Three potential approaches are discussed here. First, the federal government could continue to allow industry to self-regulate, especially where technology evolves quickly. This may promote innovative space, but relies on the industry to exercise responsible and accountable practices.

Second, Congress and the Administration could maintain current regulatory jurisdiction but leverage a deconfliction or harmonization policy so that convergent technologies are regulated under one primary authority instead of potentially multiple authorities. Preserving existing regulatory jurisdiction may require minimal restructuring and allow relatively short timeline for implementation. While a deconfliction or harmonization policy could increase coordination, overlaying such policy on an existing regulatory framework may not present the most efficient process.

Third, the Administration could consider expanding regulatory jurisdictions and authorities to include new and emerging convergent technologies that are self-regulated. This may require a complete overhaul of the technology regulatory framework, requiring congressional action and a relatively lengthy adaptation timeline for the affected industries. Some could also view such actions as extensive regulation that stifles innovation and commercial growth. On the other hand, this approach could present an opportunity to update policies on par with technology progressions and posture for emerging capabilities.

Digital Privacy Considerations

Federal data protection laws currently in place apply to specific types of data and have varied privacy and data security provisions. A federal law that comprehensively addresses digital privacy for all types of data is not in place. While illegal use of personal information (such as identity theft and fraud) is defined and enforced by federal agencies, legal use of data generated by users or converged technologies (such as social media and IoT) is not regulated to the same extent. Transparency into the activities of legal data brokers and collectors is limited.

Congress may choose to define the role of the federal government overseeing digital privacy by introducing new comprehensive federal law(s) and/or by determining minimal required standards of digital privacy. An alternative option could be expanding existing digital privacy authorities. This could include deciding whether federal entities, such as the FTC, should have their rulemaking abilities clarified or expanded.

An expanded or new federal digital privacy policy may require a variety of decisions by Congress. Two of many potential decisions pertaining to federal digital privacy policy are determining how data privacy and data security could be addressed legislatively and determining whether various types, or categories, of personal data should be treated equally or differently under varied guidance.

Data Security Considerations

Data security, as it pertains to technological convergence, may impact both the cyber and physical fronts. Some of the federal data protection laws currently in place have data security provisions, though they vary and may be focused predominantly on the cyber-aspect. This also means that different data security protocols apply to different types of data. For instance, the guidance for notifying users when personal data gets compromised is different for health, financial, and location data.

Similar to the digital privacy considerations, Congress could begin by determining whether overarching legislation for data security is necessary. Congress may consider new legislation explicitly addressing data security concerns pertaining to technological convergence. Or, Congress may consider new legislation to expand existing cybersecurity missions to address data security issues. Data security is often considered as a component of cybersecurity, but protection of the data is equally important as safeguarding a network or a system.

As with any security challenge, finding the right balance between convenience and security measures is a key component of an effective security policy. A data security policy that predominantly focuses on security measures to address potential vulnerabilities created by converged technologies could negate convenient features and beneficial capabilities, such as ubiquitous access, offered by the converged technologies. On the other hand, allowing maximum accessibility without a security measure exposes both the data and the system to risks. Not having an updated data security policy relies on existing cybersecurity measures to address potential vulnerabilities introduced by technological convergence.

Congress may determine whether data privacy and data security should be addressed in one policy. Data privacy and data security are linked and complementary, especially for digital information. While two coupled topics could be addressed in a single policy, data privacy and data security are two distinct issues. Having separate complementary policies could potentially focus more clearly on specific aspects of each issue.

In: Key Congressional Reports for May 2019 ISBN: 978-1-53616-335-3
Editor: Aston Dunkley

Chapter 2

DIGITAL TRADE AND U.S. TRADE POLICY (UPDATED)*

Rachel F. Fefer, Shayerah Ilias Akhtar and Wayne M. Morrison

ABSTRACT

As the global internet develops and evolves, digital trade has become more prominent on the global trade and economic policy agenda. The economic impact of the internet was estimated to be $4.2 trillion in 2016, making it the equivalent of the fifth-largest national economy. The digital economy accounted for 6.9% of current-dollar gross U.S. domestic product (GDP) in 2017. Digital trade has been growing faster than traditional trade in goods and services.

Congress has an important role to play in shaping global digital trade policy, from oversight of agencies charged with regulating cross-border data flows to shaping and considering legislation implementing new trade rules and disciplines through trade negotiations. Congress also works with the executive branch to identify the right balance between digital trade and other policy objectives, including privacy and national security.

* This is an edited, reformatted and augmented version of Congressional Research Service, Publication No. R44565, dated May 21, 2019.

Digital trade includes end-products, such as downloaded movies, and products and services that rely on or facilitate digital trade, such as productivity-enhancing tools like cloud data storage and email. In 2017, U.S. exports of information and communications technology-enabled services (excluding digital goods) were an estimated $439 billion. Digital trade is growing on a global basis, contributing more to global domestic product (GDP) than financial or merchandise flows.

The increase in digital trade raises new challenges in U.S. trade policy, including how to best address new and emerging trade barriers. As with traditional trade barriers, digital trade constraints can be classified as tariff or nontariff barriers. In addition to high tariffs, barriers to digital trade may include localization requirements, cross border data flow limitations, intellectual property rights (IPR) infringement, forced technology transfer, web filtering, economic espionage, and cybercrime exposure or state-directed theft of trade secrets. China's policies, in particular, such as those on internet sovereignty and cybersecurity, pose challenges for U.S. companies.

Digital trade issues often overlap and cut across policy areas, such as IPR and national security; this raises questions for Congress as it weighs different policy objectives. The Organisation for Economic Co-operation and Development (OECD) points out three potentially conflicting policy goals in the internet economy: (1) enabling the internet; (2) boosting or preserving competition within and outside the internet; and (3) protecting privacy and consumers, more generally.

While no multilateral agreement on digital trade exists in the World Trade Organization (WTO), other WTO agreements cover some aspects of digital trade. Recent bilateral and plurilateral agreements have begun to address digital trade rules and barriers more explicitly. For example, the proposed U.S.-Mexico-Canada Agreement (USMCA) and ongoing plurilateral discussions in the WTO on a potential e-commerce agreement could address digital trade barriers to varying degrees. Digital trade is also being discussed in a variety of international forums, providing the United States with multiple opportunities to engage in and shape global norms.

With workers in the high-tech sector in every U.S. state and congressional district, and over two-thirds of U.S. jobs requiring digital skills, Congress has an interest in ensuring and developing the global rules and norms of the internet economy in line with U.S. laws and norms, and in establishing a U.S. trade policy on digital trade that advances U.S. interests.

INTRODUCTION

The rapid growth of digital technologies in recent years has created new opportunities for U.S. consumers and businesses but also new challenges in international trade. For example, consumers today access e-commerce, social media, telemedicine, and other offerings not imagined thirty years ago. Businesses use advanced technology to reach new markets, track global supply chains, analyze big data, and create new products and services. New technologies facilitate economic activity but also create new trade policy questions and concerns. Data and data flows form a pillar of innovation and economic growth.

The "digital economy" accounted for 6.9% of U.S. GDP in 2017, including (1) information and communications technologies (ICT) sector and underlying infrastructure, (2) digital transactions or e-commerce, and (3) digital content or media.[1] The digital economy supported 5.1 million jobs, or 3.3% of total U.S. employment in 2017, and almost two-thirds of jobs created in the United States since 2010 required medium or advanced levels of digital skills.[2] As digital information increases in importance in the U.S. economy, issues related to digital trade have become of growing interest to Congress.

While there is no globally accepted definition of digital trade, the U.S. International Trade Commission (USITC) broadly defines digital trade as follows:

> The delivery of products and services over the Internet by firms in any industry sector, and of associated products such as smartphones and Internet-connected sensors. While it includes provision of e-commerce platforms and related services, it excludes the value of sales of physical goods ordered online, as well as physical goods that have a digital

---

[1] U.S. Bureau of Economic Analysis, *Measuring the Digital Economy: An Update Incorporating Data from the 2018 Comprehensive Update of the Industry Economic Accounts*, March 2018, https://www.bea.gov/research/papers/2018/ defining-and-measuring-digital-economy. Note: BEA did not include partially digital items, such as sharing economy services, in its estimates.

[2] Penny Pritzker and John Engler, Director Edward Alden, *The Work Ahead: Machines, Skills, and U.S. Leadership in the Twenty-First Century, Independent Task Force Report*, The Council for Foreign Relations, April 2018.

counterpart (such as books, movies, music, and software sold on CDs or DVDs).[3]

The rules governing digital trade are evolving as governments across the globe experiment with different approaches and consider diverse policy priorities and objectives. Barriers to digital trade, such as infringement of intellectual property rights (IPR) or protective industrial policies, often overlap and cut across sectors. In some cases, policymakers may struggle to balance digital trade objectives with other legitimate policy issues related to national security and privacy. Digital trade policy issues have been in the spotlight recently, due in part to the rise of new trade barriers, heightened concerns over data privacy, and an increasing number of cybertheft incidents that have affected U.S. consumers and companies. These concerns may raise the general U.S. interest in promoting, or restricting, cross-border data flows and in enforcing compliance with existing rules. Congress has an interest in ensuring the global rules and norms of the internet economy are in line with U.S. laws and norms.

Trade negotiators continue to explore ways to address evolving digital issues in trade agreements, including in the proposed U.S.-Mexico-Canada Agreement (USMCA). Congress has an important role in shaping digital trade policy, including oversight of agencies charged with regulating cross-border data flows, as part of trade negotiations, and in working with the executive branch to identify the right balance between digital trade and other policy objectives.

This chapter discusses the role of digital trade in the U.S. economy, barriers to digital trade, digital trade agreement provisions and negotiations, and other selected policy issues.

[3] Ibid. and U.S. International Trade Commission, *Global Digital Trade 1: Market Opportunities and Key Foreign Trade Restrictions*, August 2017, p.33, https://www.usitc.gov/publications/332/pub4716.pdf.

ROLE OF DIGITAL TRADE IN THE U.S. AND GLOBAL ECONOMY

The internet is not only a facilitator of international trade in goods and services, but is itself a platform for new digitally-originated services. The internet is enabling technological shifts that are transforming businesses. According to one estimate, the volume of global data flows (sending of digital data such as from streaming video, monitoring machine operations, sending communications) is growing faster than trade or financial flows. One analysis forecasts the global flows of goods, foreign direct investment (FDI), and digital data will add 3.1% to gross domestic product (GDP) from 2015-2020. The volume of global data flows is growing faster than trade or financial flows, and its positive GDP contribution offsets the lower growth rates of trade and FDI (see Figure 1).[4] Focusing domestically, the Bureau of Economic Analysis (BEA) estimates that, from 1997-2017, real value added for the digital economy outpaced overall growth in the economy each year and, in 2017, the real value-added growth of the digital economy accounted for 25% of total real GDP growth.[5]

The increase in the digital economy and digital trade parallels the growth in internet usage globally. According to one study, over half of the world's population use the internet, including 95% of people in North America.[6] As of 2017, 75% of U.S. households use wired internet access, but an increasing number rely on mobile internet access as the internet is integrated into people's everyday lives; 72% of U.S. adults own a smartphone.[7] As of the end of 2018, approximately 40% of internet traffic

---

[4] Gary Clyde Hufbauer and Zhiyao (Lucy) Lu, "Can Digital Flows Compensate for Lethargic Trade and Investment?" Peterson Institute of International Economics, November 28, 2018.

[5] U.S. Bureau of Economic Analysis, *Measuring the Digital Economy: An Update Incorporating Data from the 2018 Comprehensive Update of the Industry Economic Accounts*, March 2018, p. 6.

[6] Internet World Stats, *World Internet Usage and Population Statistics*, June 30, 2018, https://internetworldstats.com/ stats.htm.

[7] U.S. International Trade Commission, Global Digital Trade 1: Market Opportunities and Key Foreign Trade Restrictions, Publication Number: 4716, Investigation Number: 332-561, August 2017, p.47-49, https://www.usitc.gov/ publications/332/pub4716.pdf.

in the United States came from mobile devices.[8] Each day, companies and individuals across the United States depend on the internet to communicate and transmit data via various media and channels that continue to expand with new innovations (see Figure 2).

Cross-border data and communication flows are part of digital trade; they also facilitate trade and the flows of goods, services, people, and finance, which together are the drivers of globalization and interconnectedness. The highest levels reportedly are those flows between the United States and Western Europe, Latin America, and China. Efforts to impede cross-border data flows could decrease efficiency and other potential benefits of digital trade.

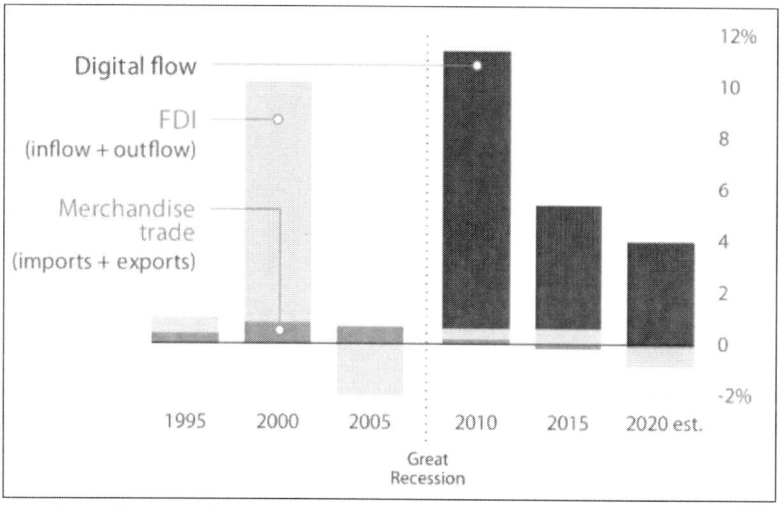

Source: Gary Clyde Hufbauer and Zhiyao Lu, "Can Digital Flows Compensate for Lethargic Trade and Investment?," Peterson Institute for International Economics, November 28, 2018.

Notes: Global internet traffic, measured in petabyte per month. Merchandise trade and FDI are normalized by dividing flows by world GDP; data flow is normalized by dividing flows by world population.

Figure 1. Effect on World GDP (percent).

[8] Statistica, "Percentage of mobile device website traffic in the United States from 1st quarter 2015 to 4th quarter 2018," 2019, https://www.statista.com/statistics/683082/share-of-website-traffic-coming-from-mobile-devices-usa/.

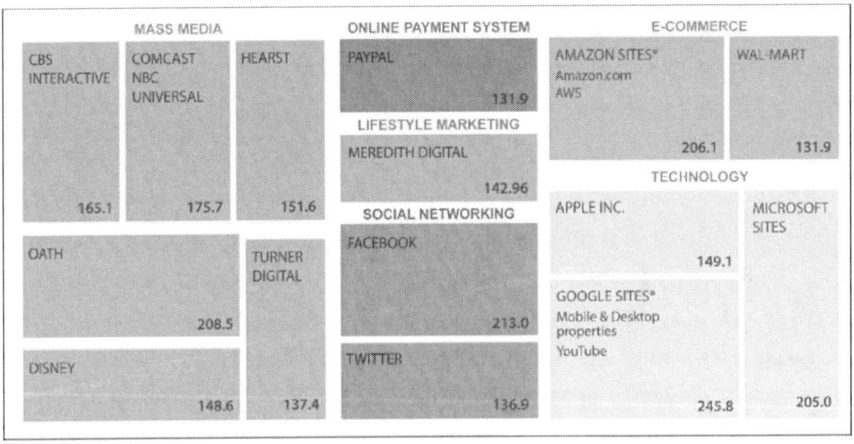

Source: Statistica.com.

Note: * Examples of web properties owned by multinational companies.

Figure 2. Snapshot of Most Popular Websites December 2018, Millions of unique U.S. visitors.

Powering all these connections and data flows are underlying ICT.[9] ICT spending is a large and growing component of the international economy and essential to digital trade and innovation.

According to the United Nations, world trade in ICT physical goods grew to $2 trillion in 2017 with U.S. ICT goods exports over $146 billion.[10]

Semiconductors, a key component in many electronic devices, are a top U.S. ICT export. Global sales of semiconductors were $468.8 billion in 2018, an increase of 6.81% over the prior year.[11] U.S.-based firms have the largest global market share with 45% and accounted for 47.5% of the Chinese market. Given the importance of semiconductors to the digital economy and continued advances in innovation, countries such as China are seeking to grow their own semiconductor industry to lessen their dependence on U.S. exports.

[9] ICT is an umbrella term that includes any communication device or application, including radio, television, cellular phones, computer and network hardware and software, satellite systems, and associated services and applications.

[10] https://unctadstat.unctad.org/wds/TableViewer/tableView.aspx?ReportId=15850.

[11] Semiconductor Industry Association, 2019 SIA Factbook, 2019.

ICT services are outpacing the growth of international trade in ICT goods. The OECD estimates that ICT services trade increased 40% from 2010 to 2016. The United States is the fourth-largest OECD exporter of ICT services, after Ireland, India, and the Netherlands.[12] ICT services include telecommunications and computer services, as well as charges for the use of intellectual property (e.g., licenses and rights). ICT-enabled services are those services with outputs delivered remotely over ICT networks, such as online banking or education. ICT services can augment the productivity and competitiveness of goods and services. In 2017, exports of ICT services grew to $71 billion of U.S. exports while services exports that could be ICT-enabled were another $439 billion, demonstrating the impact of the internet and digital revolution.[13]

ICT and other online services depend on software; the value added to U.S. GDP from support services and software has increased over the past decade relative to that of telecommunications and hardware.[14] According to one estimate, software contributed more than $1.14 trillion to the U.S. value added to GDP in 2016, an increase of 6.4% over 2014, and the U.S. software industry accounted for 2.9 million jobs directly in 2016.[15] Internet-advertising, an industry that would not exist without ICT, generated an additional 10.4 million U.S. jobs.[16]

---

[12] OECD (2017), OECD Digital Economy Outlook 2017, OECD Publishing, Paris, http://dx.doi.org/10.1787/ 9789264276284-en.

[13] According to the Department of Commerce, potentially-ICT enabled services are those that "can predominantly be delivered remotely over ICT networks, a subset of which are actually delivered via that method" and U.S. Bureau of Economic Analysis (BEA), Table 3.1. U.S. Trade in ICT and Potentially ICT-Enabled Services, by Type of Service October 19, 2018.

[14] U.S. Bureau of Economic Analysis, *Measuring the Digital Economy: An Update Incorporating Data from the 2018 Comprehensive Update of the Industry Economic Accounts*, March 2018, p.9.

[15] EIU estimates, "The Growing $1 Trillion Economic Impact of Software," software.org.

[16] John Deighton, "Economic Value of the Advertising-Supported Internet Ecosystem," 2017, https://www.iab.com/wpcontent/uploads/2017/03/Economic-Value-Study-2017-FINAL2. pdf.

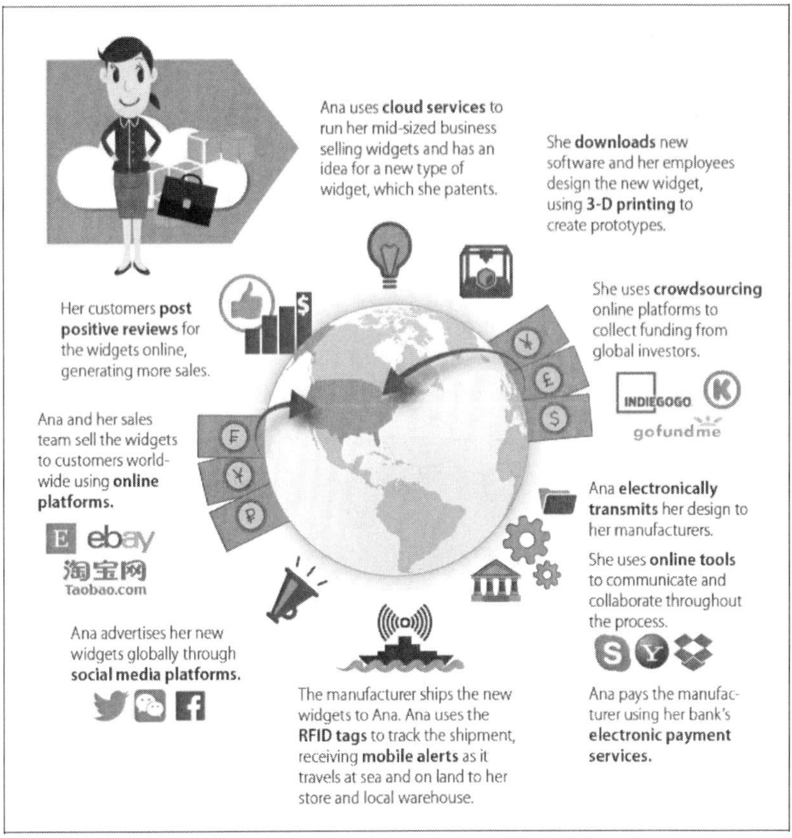

Source: CRS.

Note: The above graphic is illustrative only and is not based on a real business or reflective of all aspects of digital trade.

Figure 3.What is Digital Trade? Examples of international digital trade.

Economic Impact of Digital Trade

As the internet and technology continue to develop rapidly, increasing digitization affects finance and data flows, as well as the movement of goods and people. Beyond simple communication, digital technologies can affect global trade flows in multiple ways and have broad economic impact (see Figure 3). First, digital technology enables the creation of new goods and services, such as e-books, online education, or online banking services.

Digital technologies may also add value by raising productivity and/or lowering the costs and barriers related to flows of traditional goods and services. For example, companies may rely on radio-frequency identification (RFID) tags for supply chain tracking, 3-D printing based on data files, or devices or objects connected via the Internet of Things (see text box). In addition, digital platforms serve as intermediaries for multiple forms of digital trade, including e-commerce, social media, and cloud computing. In these ways, digitization pervades every industry sector, creating challenges and opportunities for established and new players.

Looking at digital trade in an international context, approximately 12% of physical goods are traded via international e-commerce.[17] Global e-commerce grew from $19.3 trillion in 2012 to $27.7 trillion in 2016, of which 86% was business-to-business (B2B).[18] One source estimates that cross-border business-to-consumer (B2C) e-commerce sales will reach approximately $1 trillion by 2020.[19]

These estimates do not quantify the additional benefits of digitization upon business efficiency and productivity, or of increased customer and market access, which enable greater volumes of international trade for firms in all sectors of the economy. Digitization efficiencies have the potential to both increase and decrease international trade. For example, one analysis found that logistics optimization technologies could reduce shipping and customs processing times by 16% to 28%, boosting overall trade by 6% to 11% by 2030; at the same time, however, automation, Artificial Intelligence (AI), and 3-D printing could enable more local production, thereby reducing global trade by as much as 10% by 2030.[20] The overall impact of digitization has yet to be seen.

One study coined the term "digital spillovers" to fully capture the digital economy and estimated the global digital economy, including such

---

[17] Jacques Bughin and Susan Lund, "The ascendancy of international data flows," *VOX*, January 9, 2017.

[18] U.S. International Trade Commission, Global Digital Trade 1: Market Opportunities and Key Foreign Trade Restrictions, Publication Number: 4716, Investigation Number: 332-561, August 2017, p.13, https://www.usitc.gov/ publications/332/pub4716.pdf.

[19] Susan Lund et al., "Globalization in transition: The future of trade and value chains," *McKinsey*, January 2019.

[20] Ibid.

spillovers, was $11.5 trillion in 2016, or 15.5% of global GDP.[21] Their analysis indicated that the long-term return on investment (ROI) for digital technologies is 6.7 times that of nondigital investments.[22]

Key Emerging Technologies

Internet of Things (IoT)

"encompass(es) all devices and objects whose state can be read or altered via the internet, with or without the active involvement of individuals.... The internet of things consists of a series of components of equal importance—machine-to-machine communication, cloud computing, big data analysis, and sensors and actuators. Their combination, however, engenders machine learning, remote control, and eventually autonomous machines and systems, which will learn to adapt and optimise themselves."[23]

Blockchain

"is a distributed record-keeping system (each user can keep a copy of the records) that provides for auditable transactions and secures those transactions with encryption. Using blockchain, each transaction is traceable to a user, each set of transactions is verifiable, and the data in the blockchain cannot be edited without each user's knowledge. Compared to traditional technologies, blockchain allows two or more parties without a trusted relationship to engage in reliable transactions without relying on intermediaries or central authority (e.g., a bank or government)."[24]

Artificial Intelligence (AI)

"AI can generally be thought of as computerized systems that work and react in ways commonly thought to require intelligence, such as solving complex problems in real-world situations."[25]

---

[21] Huawei Technologies and Oxford Economics, *Digital Spillover*, http://www.huawei.com/minisite/gci/en/digitalspillover/files/gci_digital_spillover.pdf.

[22] Ibid.

[23] OECD (2015), *OECD Digital Economy Outlook 2015*, p. 61, OECD Publishing, Paris. DOI: http://dx.doi.org/ 10.1787/9789264232440-2-en.

[24] For more information, see CRS In Focus IF10810, *Blockchain and International Trade*, by Rachel F. Fefer.

[25] For more information, see CRS In Focus IF10608, *Overview of Artificial Intelligence*, by Laurie A. Harris.

Blockchain is one emerging software technology some companies are using to increase efficiency and transparency and lower supply chain costs that depends on open data flows of digital trade.[26] For example, in an effort to streamline processes, save costs, and improve public health outcomes, Walmart and IBM built a blockchain platform to increase transparency of global supply chains and improve traceability for certain imported food products.[27] The initiative aims to expand to include several multinational food suppliers, farmers, and retailers and depends on connections via the Internet of Things and open international data flows. With increased applications, the Internet of Things may have a global economic impact of as much as $11.1 trillion per year, according to one study.[28]

Because of its ubiquity, the benefits and economic impact of digitization are not restricted to certain geographic areas, and businesses and communities in every U.S. state feel the impact of digitization as new business models and jobs are created and existing ones disrupted.[29] One study found that the more intensively a company uses the internet, the greater the productivity gain. The increase in internet usage is also associated with increased value and diversity of products being sold.[30]

The internet, and cloud services specifically, has been called the great equalizer, since it allows small companies access to the same information and the same computing power as large firms using a flexible, scalable, and on-demand model. For example, Thomas Publishing Co., a U.S. mid-sized, private, family-owned and -operated business, is transporting data from its own computer servers to data centers run by Amazon.com Inc.[31] Digital platforms can minimize costs and enable small and medium-sized

---

[26] For more on blockchain, see CRS Report R45116, *Blockchain: Background and Policy Issues*, by Chris Jaikaran.

[27] Roger Aitken, "IBM & Walmart Launching Blockchain Food Safety Alliance In China With Fortune 500's JD.com," *Forbes*, December 14, 2017.

[28] Alexandre Menard, "How can we recognize the real power of the Internet of Things?" *McKinsey*, November 2017.

[29] John Wu, Adams Nager, and Joseph Chuzhin, *High-Tech Nation: How Technological Innovation Shapes America's 435 Congressional Districts*, ITIF, November 28, 2016, p. 4, https://itif.org/publications/2016/11/28/technation.

[30] The World Bank Group, *World Development Report 2016: Digital Dividends*, 2016, http://www.worldbank.org/en/ publication/wdr2016.

[31] Jay Greene, "Amazon to Launch Cloud Migration Service," *The Wall Street Journal*, March 15, 2016.

enterprises (SMEs) to grow through extended reach to customers or suppliers or integrating into a global value chain (GVC). More than 50% of businesses globally rely on data flows for cloud computing (see text box).[32]

Digitization of customs and border control mechanisms also helps simplify and speed delivery of goods to customers. Regulators are looking to blockchain technology to improve efficiency in managing and sharing data for functions such as border control and customs processing of international shipments.[33] With simpler border and customs processes, more firms are able to conduct business in global markets (or are more willing to do so). A study of U.S. SMEs on the e-commerce platform eBay found that 97% export, while that number is a full 100% in countries as diverse as Peru and Ukraine.[34] Netflix, a U.S. firm offering online streaming services, increased its international revenue from $4 million in 2010 to more than $5 billion in 2017.[35]

A Local Manufacturer Grows Through Digital Trade

Kirk Anton and Tricia Hudson launched their business in 2010 as a one-stop shop for heat transfer materials, importing heat-applied materials and creating customized products for clients. The company moved online and converted to completely digital in 2013, using services such as Google Analytics to inform their marketing campaigns. Growing by more than 70% over four years, in 2017 the company employed forty people in Florida, Kentucky, Nevada, and North Dakota, and boasted over 85,000 customers across the globe.[36]

---

[32] U.S. International Trade Commission, *Global Digital Trade 1: Market Opportunities and Key Foreign Trade Restrictions*, Investigation Number: 332-561, August 2017.

[33] Commercial Customs Operations Advisory Committee (COAC), *Trade Progress Report*, November 2017, https://www.cbp.gov/sites/default/files/assets/documents/2017-Nov/Global%20Supply%20Chain%20Subcommittee%20Trade%20Executive%20Summary%20Nov%202017.pdf.

[34] James Manyika, Sree Ramaswamy, and Somesh Khanna et al., *Digital America: A Tale of the Haves and Have-Mores*, McKinsey Global Institute, December 2015, p. 40, http://www.mckinsey.com/industries/high-tech/our-insights/ digital-america-a-tale-of-the-haves-and-have-mores.

[35] World Trade Organization, "World Trade Report 2018: The future of world trade," p.10, 2018, https://www.wto.org/ english/res_e/publications_e/wtr18_e.htm.

[36] Google, *Economic Impact United States 2017*, https://economicimpact.google.com/.

A similar argument has been made for firms and governments in low- and middle-income countries who can take advantage of the power of the internet to foster economic development. According to one official of the Asia-Pacific Economic Cooperation Forum (APEC), technology has enabled SMEs to open in new sectors such as ride-sharing and online order delivery services, and provides them with a "bigger, better opportunity to grow and learn that to join a global value chain."[37] Another study of SMEs estimated that the internet is a net creator of jobs, with 2.6 jobs created for every job that may be displaced by internet technologies; companies that use the internet intensively effectively doubled the average number of jobs.[38] However, the costs of digital trade can be concentrated on particular sectors (see next section).

Digitization Challenges

The U.S. digital economy supported 3.3% of total U.S. employment in 2017, and those jobs earned approximately one and a half times the average annual worker compensation of the overall U.S. economy, making them attractive source for future growth.[39] Software, and the software industry, contributes to the GDP in all 50 states, with the value-added GDP of the software industry growing more than 40% in Idaho and North Carolina.[40] Industries, such as media and firms in urban centers, account for a larger share of the benefits. Many in business and research communities are only beginning to understand how to take advantage of the vast amounts of data being collected every day.

However, sources of "e-friction" or obstacles can prevent consumers, companies, and countries from realizing the full benefits of the online

---

[37] APEC, "APEC's Startup Revolution Brings the Next Big Thing," November 2, 2017; https://www.apec.org/Press/ Features/2017/1102_interview.

[38] Matthieu Pélissié du Rausas, James Manyika, and Eric Hazan et al., *Internet matters: The Net's sweeping impact on growth, jobs, and prosperity*, McKinsey Global Institute, May 2011, p. 21, http://www.mckinsey.com/industries/hightech/our-insights/internet-matters.

[39] U.S. Bureau of Economic Analysis, *Measuring the Digital Economy: An Update Incorporating Data from the 2018 Comprehensive Update of the Industry Economic Accounts*, March 2018, p.2.

[40] Software.org, "The Growing $1 Trillion Economic Impact of Software," https://software.org/ reports/2017-ussoftware-impact/.

economy.[41] Causes of e-friction can fall into four categories: infrastructure, industry, individual, and information. Government policy can influence e-friction, from investment in infrastructure and education to regulation and online content filtering. According to some experts, economies with lower amounts of e-friction may be associated with larger digital economies.[42]

While there are numerous positive digital dividends, there are also possible negative and uneven results across populations, such as the displacement of unskilled workers, an imbalance between companies with and without internet access, and the potential for some to use the internet to establish monopolies.[43] While new technologies and new business models present opportunities to enhance efficiency and expand revenues, innovate faster, develop new markets, and achieve other benefits, new challenges also arise with the disruption of supply chains, labor markets, and some industries. For example, one study found a mismatch between workforce skills and job openings such as in Nashville, TN, which has an abundance of workers with music production and radio broadcasting skills but a scarcity of workers with IT infrastructure, systems management, and web programming skills.[44] Another source notes over 11,000 open computing jobs in Michigan, with average salaries of over $80,000.[45]

The World Bank identified policy areas to try to ensure, and maintain, the potential benefits of digitization. Policy areas include establishing a favorable and competitive business climate, developing strong human capital, ensuring good governance, investing to improve both physical and digital infrastructure, and raising digital literacy skills. According to the World Economic Forum Global Competitiveness Index 4.0, the United States is ranked at the top with a score of 85.6% compared to the global

---

41 Paul Zwillenberg, Dominic Field, and David Dean, *Greasing the Wheels of the Internet Economy*, Boston Consulting Group, February 2014, https://www.bcgperspectives.com/content/articles/digital_economy_telecommunications_greasing_wheels_internet_economy/.

42 Ibid.

43 The World Bank Group, *World Development Report 2016: Digital Dividends*, 2016, http://www.worldbank.org/en/ publication/wdr2016.

44 Penny Pritzker and John Engler, Director Edward Alden, *The Work Ahead: Machines, Skills, and U.S. Leadership in the Twenty-First Century, Independent Task Force Report*, Council of Foreign Relations, April 2018.

45 https://code.org/promote/mi.

median score of 60%.[46]The study identifies the key drivers of productivity as human capital, innovation, resilience, and agility, noting that future productivity depends not only on investment in technology but investment in digital skills. While the United States is considered a "super innovator," the report also notes "indications of a weakening social fabric ... and worsening security situation ... as well as relatively low checks and balances, judicial independence, and transparency."[47]

With the rapid pace of technology innovation, more jobs may become automated, with digital skills becoming a foundation for economic growth for individual workers, companies, and national GDP.[48] Over two-thirds of U.S. jobs created since 2010 require some level of digital skills.[49] The OECD found that generic ICT skills are insufficient among a significant percentageof the global workforce and few countries have adopted comprehensive ICT skills strategies to help workers adapt to changing jobs.[50]

DIGITAL TRADE POLICY AND BARRIERS

Policies that affect digitization in any one country's economy can have consequences beyond its borders, and because the internet is a global "network of networks," the state of a country's digital economy can have global ramifications. Protectionist policies may erect barriers to digital trade, or damage trust in the underlying digital economy, and can result in the fracturing, or so-called balkanization, of the internet, lessening any gains. What some policymakers see as protectionist, however, others may view as necessary to protect domestic interests. For examples of the types

---

[46] World Economic Forum, Global Competitiveness Report 2018, p. 10, http://www3.weforum.org/docs/GCR2018/05FullReport/TheGlobalCompetitivenessReport2018.pdf.

[47] Ibid, p. 33.

[48] The World Bank Group, *World Development Report 2016: Digital Dividends*, 2016, http://www.worldbank.org/en/ publication/wdr2016.

[49] Penny Pritzker and John Engler, Director Edward Alden, *The Work Ahead: Machines, Skills, and U.S. Leadership in the Twenty-First Century, Independent Task Force Report*, Council of Foreign Relations, April 2018.

[50] OECD (2017), OECD Digital Economy Outlook 2017, OECD Publishing, Paris, http://dx.doi.org/10.1787/ 9789264276284-en.

of digital trade barriers that are in place around the globe, please see Appendix.

Despite common core principles such as protecting citizen's privacy and expanding economic growth, governments face multiple challenges in designing policies around digital trade. The OECD points out three potentially conflicting policy goals in the internet economy: (1) enabling the internet; (2) boosting or preserving competition within and outside the internet; and (3) protecting privacy and consumers more generally.[51]

Ensuring a free and open internet is a stated policy priority for the U.S. government.[52] Like other cross-cutting policy areas, such as cybersecurity or privacy, no one federal entity has policy primacy on all aspects of digital trade, and the United States has taken a sectoral approach to regulating digitization. According to an OECD study, the United States is the only OECD country that uses a decentralized, market-driven approach for a digital strategy rather than having an overarching national digital strategy, agenda, or program.[53]

Protect a Free and Open Internet[54]

Protecting a free and open internet is a policy priority as stated in President Trump's 2017 National Security Strategy.

"The United States will advocate for open, interoperable communications, with minimal barriers to the global exchange of information and services. The United States will promote the free flow of data and protect its interests through active engagement in key organizations, such as the Internet Corporation for Assigned Names and Numbers (ICANN), the Internet Governance Forum (IGF), the UN, and the International Telecommunication Union (ITU)."

[51] Koske, I. et al. (2014), "The Internet Economy—Regulatory Challenges and Practices," OECD Economics Department Working Papers, No. 1171, OECD Publishing, Paris. DOI, http://dx.doi.org/10.1787/5jxszm7x2qmr-en.

[52] http://www.state.gov/e/eb/cip/netfreedom/index.htm.

[53] OECD (2017), OECD Digital Economy Outlook 2017, OECD Publishing, Paris, p. 34, http://dx.doi.org/10.1787/ 9789264276284-en.

[54] The White House, *National Security Strategy of the United States of America*, December 2017, p. 41, https://www.whitehouse.gov/wp-content/uploads/2017/12/NSS-Final-12-18-2017-0905.pdf.

The Department of Commerce works to promote U.S. digital trade policies domestically and abroad. In 2015, Commerce launched a Digital Economy Agenda that identifies four pillars:[55]

1. "Promoting a free and open Internet worldwide, because the Internet functions best for our businesses and workers when data and services can flow unimpeded across borders";
2. "Promoting trust online, because security and privacy are essential if electronic commerce is to flourish";
3. "Ensuring access for workers, families, and companies, because fast broadband networks are essential to economic success in the 21st century"; and
4. "Promoting innovation, through smart intellectual property rules and by advancing the next generation of exciting new technologies."

Commerce's digital attaché program under the foreign commercial service helps U.S. businesses navigate regulatory issues and overcome trade barriers to e-commerce exports in key markets.[56]

The Administration also works to promote U.S. digital priorities by identifying and challenging foreign trade barriers and through trade negotiations. As with traditional trade barriers, digital trade constraints can be classified as tariff or nontariff barriers. Tariff barriers may be imposed on imported goods used to create ICT infrastructure that make digital trade possible or on the products that allow users to connect, while nontariff barriers, such as discriminatory regulations or local content rules, can block or limit different aspects of digital trade. Often, such barriers are intended to protect domestic producers and suppliers. Some estimates indicate that removing foreign barriers to digital trade could increase annual U.S. real GDP by 0.1%-0.3% ($16.7 billion-$41.4 billion), increase

---

[55] Alan B Davidson, "The Commerce Department's Digital Economy Agenda," November 9, 2015, https://www.commerce.gov/news/blog/2015/11/commerce-departments-digital-economy-agenda.

[56] For more information, see https://www.export.gov/digital-attache.

U.S. wages up to 1.4%, and add up to 400,000 U.S. jobs in certain digitally intensive industries.[57]

2015 U.S. Digital Trade Negotiating Objectives

Congress enhanced its digital trade policy objectives for U.S. trade negotiations in the Bipartisan Congressional Trade Priorities and Accountability Act of 2015 (P.L. 114-26), or Trade Promotion Authority (TPA), signed into law in June 2015.[58] TPA 2015 objectives related to digital trade direct the Administration to negotiate agreements that

- ensure application of existing WTO commitments to the digital trade environment, ensuring no less favorable
- treatment to physical trade;
- prohibit forced localization requirements and restrictions to digital trade and data flows;
- keep electronic transmissions duty-free; and
- ensure relevant legitimate regulations are as least trade restrictive as possible.

Tariff Barriers

Historically, trade policymakers focused on overt trade barriers such as tariffs on products entering countries from abroad. Tariffs at the border impact goods trade by raising the prices of products for producers or end customers, if tariff costs are passed down, thus limiting market access for U.S. exporters selling products, including ICT goods. Quotas may limit the number or value of foreign goods, persons, suppliers, or investments allowed in a market. Since 1998, WTO countries have agreed to not impose customs duties on electronic transmissions covering both goods (such as e-books and music downloads) and services.

---

[57] Digitally intensive industries include sectors in communications, finance, trade, other services, and manufacturing. U.S. International Trade Commission, *Digital Trade in the U.S. and Global Economies, Part 2*, Publication No: 4485, Investigation No: 332-540, August 2014, pp. 106-108, https://www.usitc.gov/publications/332/pub4485.pdf.

[58] For more information on TPA, see CRS In Focus IF10038, *Trade Promotion Authority (TPA)*, by Ian F. Fergusson, and CRS Report RL33743, *Trade Promotion Authority (TPA) and the Role of Congress in Trade Policy*, by Ian F. Fergusson.

While the United States is a major exporter and importer of ICT goods, tariffs are not levied on many of the products due to free trade agreements (FTAs) and the World Trade Organization (WTO) Information Technology Agreement (see below). Tariffs may still serve as trade barriers for those countries or products not covered by existing FTAs or the WTO ITA.

U.S. ICT services are often inputs to final demand products that may be exported by other countries, such as China. U.S. ICT services have shown increasing growth rates since the middle of 2014.[59]

ICT Goods Tariff Barriers: Selected Examples

Brazil, Mexico, and Vietnam are key participants in the ICT goods market and impose high tariffs on non-FTA partners. According to the United Nations Statistics Division, in 2015 Brazil reported $1.3 billion in medical ICT equipment imports, such as electrocardiographs, ultrasound devices, and magnetic resonance imaging devices,[60] despite tariffs of up to 16% on these products.[61]

In 2014, Vietnam reportedly imported $10.3 billion worth of electronic integrated circuits (microchips) and parts, including approximately 4% or $398 million from the United States.[62] While Vietnam imposes no tariffs on these product categories, several ICT items in Vietnam's tariff schedule have high applied rates, including multiple categories of radio equipment, which have an applied rate as high as 30% according to the WTO.[63]

Nontariff Barriers

Nontariff barriers (NTBs) are not as easily quantifiable as tariffs. Like digital trade, NTBs have evolved and may pose significant hurdles to companies seeking to do business abroad. NTBs often come in the form of laws or regulations that intentionally or unintentionally discriminate and/or hamper the free flow of digital trade.

---

59 OECD (2017), OECD Digital Economy Outlook 2017, OECD Publishing, Paris, p. 120, http://dx.doi.org/10.1787/ 9789264276284-en.

60 Data on Harmonized System code 9018 from U.N. Comtrade: http://comtrade.un.org.

61 CRS analysis of tariff data from the WTO Tariff Analysis Online (TAO): https://tao.wto.org.

62 U.S. Census Bureau.

63 Harmonized System code 8527, from WTO TAO.

Nondiscrimination between local and foreign suppliers is a core principle encompassed in global trading rules and U.S. free trade agreements. While WTO agreements cover physical goods, services, and intellectual property, there is no explicit provision for nondiscrimination for digital goods. As such, NTBs that do not treat digital goods the same as physical ones could limit a provider's ability to enter a market.

Potential Barriers to Digital Trade

- High tariffs
- Localization requirements
- Cross border data flow limitations
- IPR infringement
- Discriminatory, unique standards or burdensome testing
- Filtering or blocking
- Restrictions on electronic payment systems or the use of encryption
- Cybertheft of U.S. trade secrets
- Forced technology transfer

Broader governance issues, including rule of law, transparency, and investor protections, can pose barriers and limit the ability of firms and individuals to successfully engage in digital trade. Similarly, market access restrictions on investment and foreign ownership, or on the movement of people, whether or not specific to digital trade or ICT sectors, may limit a company's ability enter a foreign market. Other NTBs are more specific to digital trade.

Localization Requirements

Localization measures are defined as measures that compel companies to conduct certain digital trade-related activities within a country's borders.[64] Governments often use privacy protection or national security arguments as justifications for these measures. Though localization policies can be used to achieve legitimate public policy objectives, some

[64] U.S. International Trade Commission, *Digital Trade in the U.S. and Global Economies, Part 1*, Publication No: 4415, Investigation No: 332-531, July 2013, p. 16, https://www.usitc.gov/publications/332/pub4415.pdf.

are designed to protect, favor, or stimulate domestic industries, service providers, or intellectual property at the expense of foreign counterparts and, in doing so, function as nontariff barriers to market access. In recent free trade agreements, the United States has aimed to ensure an open internet and eliminate digital trade barriers, while preserving flexibility for governments to pursue legitimate policy objectives (see below).

Cross-Border Data Flow Restrictions

According to a 2017 USITC report, data localization was the most cited policy measure impeding digital trade, and the number of data localization measures globally has doubled in the last six years.[65] One study found that over 120 countries have laws related to personal data protection, often requiring data localization.[66] Regulations limiting cross-border data flows and requiring local storage are a type of localization requirement that prohibit companies from exporting data outside a country.

Such restrictions can pose barriers to companies whose transactions rely on the internet to serve customers abroad and operate more efficiently. For example, data localization requirements can limit e-commerce transactions that depend on foreign financial service providers or multinational firms' full analysis of big data from across an entire company or global value chain. Regulations limiting cross-border data flows may force companies to build local server infrastructure within a country, not only increasing costs and decreasing scale, but also creating data silos that may be more vulnerable to cybersecurity risks. According to some analysts, computing costs in markets with localization measures can be 30%-60% higher than in more open markets.[67]

Data localization requirements pose barriers to companies' efforts to operate more efficiently by migrating to the cloud or to SMEs attempting to enter new markets. According to some estimates, cloud computing

[65] https://www.usitc.gov/publications/332/pub4716.pdf.

[66] C&M International, "Benefits of the APEC Cross-Border Privacy Rules," October 2018, https://www.crowell.com/files/20181001-Benefits-of-CBPR-System%20Guide_Oct%202018_final.pdf.

[67] David J. Lynch, "The U.S. dominates the world of big data. But Trump's NAFTA demands could put that at risk.," *Washington Post*, November 28, 2018.

accounted for 70% of related IT market growth between 2012 and 2015, and is expected to represent 60% of growth through 2020.[68] Most of the largest global providers of cloud computing services are U.S. companies (Amazon, Microsoft, Google, and IBM).

Regulations or policies that limit data flows create barriers to firms and countries seeking to consume cloud services. One U.S. business group noted increased forced localization measures, citing examples in China, Colombia, the European Union (EU), Indonesia, South Korea, Russia, and Vietnam.[69] The Business Software Alliance's 2018 Global Cloud Computing Scorecard highlighted barriers to cloud services in Indonesia, Russia, and Vietnam.[70] For example, to comply with localization requirements and continue to serve consumers of Google's many cloud services (e.g., Gmail, search, maps) globally, the company is opening more data centers in the United States and internationally.[71]

Finding a global consensus on how to balance open data flows, cybersecurity, and privacy protection may be key to maintaining trust in the digital environment and advancing international trade.[72] Countries are debating how to achieve the right balance and potential paths forward in plurilateral and multilateral forums and trade negotiations (see "U.S. Bilateral and Plurilateral Agreements").

Other Localization Requirements

In addition to cross-border data flow restrictions, localization policies include requirements to use local content, whether hardware or software, as a condition for manufacturing or access to government procurement

---

68 Mark Brinda and Michael Heric, "The Changing Faces of the Cloud: Technology Companies Are Adapting to Sell Cloud to the Growing Number of More-Mainstream Buyers," Bain & Company, 2017.

69 Information Technology Industry Council, Comments in Response to Executive Order Regarding Trade Agreements Violations and Abuses, August 1, 2017, http://www.itic.org/dotAsset/9d22f0e2-90cb-467d-81c8-ecc87e8dbd2b.pdf.

70 Business Software Alliance, *2018 BSA Global Cloud Computing Scorecard*, http://cloudscorecard.bsa.org/2018/pdf/ BSA_2018_Global_Cloud_Scorecard.pdf.

71 Google Cloud Platform Blog, "Google Cloud Platform adds two new regions, 10 more to come," March 22, 2016, https://cloudplatform.googleblog.com/2016/03/announcing-two-new-Cloud-Platform-Regions-and-10-more-tocome_22.html?mod=djemCIO_h.

72 For more information on data flows, see CRS Report R45584, *Data Flows, Online Privacy, and Trade Policy*, by Rachel F. Fefer.

contracts; use local infrastructure or computing facilities; or partner with a local company and transfer technology or intellectual property to that partner. Localization requirements can also pose a threat to intellectual property (discussed below).

In April 2018, the Commerce Department announced plans to develop a "comprehensive strategy to address trade-related forced localization policies, practices, and measures impacting the U.S. information and communications technology (ICT) hardware manufacturing industry."[73] In creating a strategic response to the increase in protectionist localization policies globally, Commerce aims to preserve the competitiveness of the U.S. ICT sector.[74]

Examples of Localization Barriers

Examples of localization barriers include the following:

- In China, measures across multiple sectors (e.g., banking) require "secure and controllable" technology, mandating suppliers purchase Chinese products and use Chinese suppliers (see "China").
- In Turkey, the Law on Payments and Security Settlement Systems, Payment Services and Electronic Money Institutions requires firms to maintain documents, records, data storage, and processing facilities in Turkey.
- In Nigeria, the government requires ICT companies to use Nigerian companies for the provision of at least 80% of all value-added services on their network, without clearly defining "value-added services."
- In India, the 2015 National Telecom M2M ("machine to machine") roadmap recommends preferences for locally manufactured SIM cards and domestically sourced goods.

Source: *2019 National Trade Estimate Report on Foreign Trade Barriers*, Office of the United States Trade Representative, March 2019.

Intellectual Property Rights (IPR) Infringement

While the internet and digital technologies have opened up markets for international trade, they also present ongoing and unique challenges for the

[73] Department of Commerce, "U.S. Strategy To Address Trade-Related Forced Localization Barriers Impacting the U.S. ICT Hardware Manufacturing Industry," 83 *Federal Register* 15786, April 12, 2018.

[74] The planned strategy will not address cross-border data flow restrictions.

protection and enforcement of intellectual property (IP), which are creations of the mind—such as an invention, literary/artistic work, design, symbol, name, or image—embodied in a physical or digital object. Intellectual property rights (IPR)[75] are legal, private, enforceable, time-limited rights that governments grant to inventors and artists to exclude others from using their creations without their permission. Examples of IPR include patents, copyrights, trademarks, and trade secrets.

Innovations in digital technologies fuel IPR infringement by enabling the rapid duplication and distribution of content that is low-cost and high-quality, making it easy, for instance, to pirate music, movies, software, and other copyrighted works, and to share them globally. The internet provides "ease of conducting commerce through unverified vendors, inability for consumers to inspect goods prior to purchase, and deceptive marketing."[76] Both copyright- and trademark-based industries face challenges tackling not only infringement in physical marketplaces, but increasingly also online marketplaces.[77] Cyber-enabled theft of trade secrets is of growing concern. Trade secrets are essential to many businesses' operations and important assets, including those in ICT, services, biopharmaceuticals, manufacturing, and environmental and other technologies.

IPR infringement in the digital environment is particularly difficult to quantify but considered to be significant, potentially exceeding the volume of sales through traditional physical markets.[78] A 2016 industry study estimated the value of digitally pirated music, movies, and software (not actual losses) to be $213 billion in 2013 and growing to as much as $384-$856 billion in 2022.[79] The IP Commission estimated that the annual cost to the U.S. economy from counterfeit goods, pirated software, and theft of

---

[75] See CRS Report RL34292, *Intellectual Property Rights and International Trade*, by Shayerah Ilias Akhtar and Ian F. Fergusson; and CRS In Focus IF10033, *Intellectual Property Rights (IPR) and International Trade*, by Shayerah Ilias Akhtar and Ian F. Fergusson.

[76] USTR, *2015 Out-of-Cycle Review of Notorious Markets*, December 2015, p. 9.

[77] USTR, 2017 Out-of-Cycle Review of Notorious Markets, January 2018.

[78] USTR, *2017 Special 301 Report*, April 2017.

[79] Frontier Economics, *The Economic Impacts of Counterfeiting and Piracy*, report commissioned by Business Action to Stop Counterfeiting and Piracy (BASCAP) of the International Chamber of Commerce (ICC) and the International Trademark Association (INTA), June 2017.

trade secrets continues to surpass $225 billion and could reach $600 billion.[80]

Efforts to address IPR infringement raise issues of balance about, on one hand, protecting and enforcing IPR to protect the rights of content holders and incentivize innovation in the digital environment and, on the other hand, setting appropriate limitations and exceptions to ensure other economically and socially valuable uses. Content industries say that IP theft costs them sales, detracts from legitimate services, harms investors in these businesses, damages their brand or reputation, and hurts "law-abiding" consumers. Some technology product and service companies, as well as some civil society groups, assert that overly stringent IPR policies may stifle information flows and legitimate digital trade and these groups support "fair use" exceptions and limitations to IPR.[81]

Other IPR-related barriers to digital trade include government measures, policies, and practices that are intended to promote domestic "indigenous innovation" (i.e., develop, commercialize, and purchase domestic products and technologies) but that can also disadvantage foreign companies. These measures can be linked to "forced" localization barriers to trade. China, for instance, conditions market access, government procurement, and the receipt of certain preferences or benefits on a firm's ability to show that certain IPR is developed in China or is owned by or licensed to a Chinese party. Another example is India's data and server localization requirements, which USITC firms assert hurt market access and innovation in their sector. (See below).

---

[80] The estimate does not include patent infringement. Commission on the Theft of American Intellectual Property ("IP Commission"), *The Theft of American Intellectual Property: Reassessments of the Challenge and U.S. Policy—Update to the IP Commission Report, 2017*. The IP Commission describes itself as an "independent and bipartisan initiative of leading Americans from the private sector, public service in national security and foreign affairs, academia, and politics."

[81] "Fair use" is a doctrine recognized in U.S. law that permits limited use of copyrighted works without requiring permission from the rights holder in certain cases, such as criticism, comment, news reporting, research, scholarship, and teaching.

New EU Copyright Rules

The EU's new copyright directive highlights the debate over balance, and has implications for U.S. digital trade. On April 15, 2019, the EU adopted the new rules to modernize its copyright laws to adapt to the digital environment. One objective of the directive is to create a fairer marketplace for online content for creators and press. The directive introduces an EU-wide "neighboring right" to allow news publishers to be compensated for the use of their articles by online platforms, as well provide for journalists to receive an appropriate share of the revenues generated. News platforms such as Google will have to negotiate licenses from newspapers and other publishers for showing content that is under two-years-old on their news feeds. Short extracts from press publications— sometimes called "snippets"—are outside of the scope of the rule.[82] The directive also reinforces the position of creators and right holders to negotiate and secure compensation for online use of their content hosted in the EU by major content platforms such as YouTube. If no licensing agreement exists between creators and the online platforms, YouTube and other such platforms must demonstrate "best efforts" to remove copyright materials if they are notified of infringing uploads. Newer and smaller platforms are not subject to all of these requirements. The directive addresses other digital copyright issues as well. Some U.S. stakeholders, such as the publishing industry, support the new rules, while others, including U.S. businesses that are content-aggregators, have raised concerns about increased costs, market access barriers, and effects on the innovation environment of the new rules.[83] After the publication of the directive in the Official Journal of the EU, member states will have 24 months to transpose the new rules into their national law.

National Standards and Burdensome Conformity Assessment

Local or national standards that deviate significantly from recognized international standards may make it difficult for firms to enter a particular market. An ICT product or software that conforms to international standards, for example, may not be able to connect to a local network or device based on a local or proprietary standard. Also, proprietary standards can limit a firm's ability to serve a market if their company practices or assets do not conform with (nor do their personnel have training in) those standards. As a result, U.S. companies may not be able to reach customers or partners in those countries.

---

[82] European Commission, "Question & Answers: EU Negotiators Reach a Breakthrough to Modernise Copyright Rules," press release, February 13, 2019.

[83] USTR, *2019 National Trade Estimate Report on Foreign Trade Barriers*, March 2019.

Similarly, redundant or burdensome conformity assessment or local registration and testing requirements often add time and expense for a company trying to enter a new market, and serve as a deterrent to foreign companies. For example, India's Compulsory Registration Order (CRO) mandates that manufacturers register their products with laboratories affiliated with or certified by the Bureau of Indian Standards, even if the products have already been certified by accredited international laboratories, and is an often-cited concern for U.S. businesses facing delays getting products to market.[84] If a company is required to provide the source code, proprietary algorithms, or other IP to gain market access, it may fear theft of its IP and not enter that market (see above).

Filtering, Blocking, and Net Neutrality

In some nations, government seeks strict control over digital data within its borders, such as what information people can access online, and how information is shared inside and outside its borders. Governments that filter or block websites, or otherwise impede access, form another type of nontariff barrier. For example, China has asserted a desire for "digital sovereignty" and has erected what is termed by some as the "Great Firewall." A change to China's internet filters also blocks virtual private network (or VPN) access to sites beyond the Great Firewall. VPNs have been used by Chinese citizens to use websites like Facebook and by companies to access data outside of China (e.g., information from foreign subsidiaries or partners).[85]

While China is the most well-known, it is not alone in seeking to control access to websites. For example, Thailand established a Computer Data Filtering Committee to use the court system to block websites that it views as violating public order and good order, as well as intellectual property.[86] In Russia, citizens protested government censorship, including

[84] USTR, *2019 National Trade Estimate Report on Foreign Trade Barriers*, March 2019, p. 242.
[85] Yu Nakamura, "China's war on VPNs creates havoc at foreign companies," December 17, 2017.
[86] USTR, *2018 National Trade Estimate Report on Foreign Trade Barriers*, March 2018, p. 446.

the blocking of a popular messaging application along with other websites and online tools.[87]

Several U.S. and foreign policymakers have expressed concern about the influence that violent or harmful content online may have upon those who view or read it. In response, some countries have introduced legislation to regulate internet content, for example, to fight the impact and spread of violent material and false information.[88] In the United States, significant First Amendment freedom of speech issues are raised by the prospect of government restrictions on the publication and distribution of speech, even speech that advocates terrorism.[89] As a result, what users can access online may vary across countries, depending on national policy and preferences. These differences illustrate the complexity of the internet and evolving technologies, and the lack of global standards that prevails in other areas of international trade.

National-level net neutrality policies also differ widely. Net neutrality rules govern the management of internet traffic as it passes over broadband internet access services, whether those services are fixed or wireless. Allowing internet access providers to limit or otherwise discriminate against content providers, foreign and domestic, may create a nontariff barrier.[90] In the United States, the Federal Communications Commission (FCC) classification of broadband internet service providers (ISPs) has been controversial domestically and may differ from how U.S. trading partners regulate ISPs.

Cybersecurity Risks

The growth in digital trade has raised issues related to cybersecurity, the act of protecting ICT systems and their contents from cyberattacks. Cyberattacks in general are deliberate attempts by unauthorized persons to

---

[87] Neil MacFarquhar, "'They Want to Block Our Future': Thousands Protest Russia's Internet Censorship," *The New York Times*, April 30, 2018.

[88] Adam Satariano, "Britain Proposes Broad New Powers to Regulate Internet Content," *The New York Times*, April 7, 2019.

[89] For more information, see CRS Report R44626, *The Advocacy of Terrorism on the Internet: Freedom of Speech Issues and the Material Support Statutes*, by Kathleen Ann Ruane.

[90] For more information on net neutrality, see CRS Report R40616, *The Net Neutrality Debate: Access to Broadband Networks*, by Angele A. Gilroy.

access ICT systems, usually with the goal of theft, disruption, damage, or other unlawful actions. Cybersecurity can also be an important tool in protecting privacy and preventing unauthorized surveillance or intelligence gathering.[91] Although there is overlap between data protection and privacy, the two are not equivalent. Cybersecurity measures are essential to protect data (e.g., against intrusions or theft by hackers). However, they may not be sufficient to protect privacy.

Cyberattacks can pose broad risks to financial and communication systems, national security, privacy, and digital trade and commerce. According to the White House Council of Economic Advisers, malicious cyberactivity (i.e., business disruption, theft of proprietary information) cost the U.S. economy up to $109 billion in 2016.[92] Cybersecurity risks run across all industry sectors that rely on digital information. In the entertainment industry, for example, Iranian hackers stole unreleased episodes of HBO's "Game of Thrones" series, holding them for ransom, and potentially costing the company and risking intellectual property and harm to the corporate reputation.[93] The Federal Bureau of Investigations (FBI) suspects Chinese hackers were behind a cyberattack on the Marriot's Starwood hotel chain that resulted in potentially stealing IPR and the personal information of up to 327 million hotel customers, including their birthdates and passport numbers.[94] An FBI official testified to the Senate Judiciary Committee that Chinese espionage efforts have become "the most severe counterintelligence threat facing our country today."[95]

---

[91] For more information on cybersecurity, see CRS Report R43831, *Cybersecurity Issues and Challenges: In Brief*, by Eric A. Fischer, and CRS In Focus IF10559, *Cybersecurity: An Introduction*, by Chris Jaikaran.

[92] Council of Economic Advisers, *The Cost of Malicious Cyber Activity to the U.S. Economy*, February 2018, https://www.whitehouse.gov/wp-content/uploads/2018/02/The-Cost-of-Malicious-Cyber-Activity-to-the-U.S.- Economy.pdf.

[93] Nicole Hong, "Iranian Charged With Hacking HBO, Taking 'Game of Thrones' Scripts," *Wall Street Journal*, November 21, 2017.

[94] David E. Sanger et al, "Marriott Data Breach Is Traced to Chinese Hackers as U.S. Readies Crackdown on Beijing," *The New York Times*, December 11, 2018.

[95] U.S. Congress, Senate Committee on the Judiciary, *China's Non-Traditional Espionage Against the United States: The Threat and Potential Policy Responses*, 115th Cong., December 12, 2018.

Cybersecurity threats can disrupt business operations or supply chains. The 2017 WannaCry ransomware attack impacted public and private sector entities in over 150 countries with direct costs of at least $8 billion due to computer downtime, according to one estimate.[96] In the widespread attack, computers in homes, schools, hospitals, government agencies, and companies were hit. The United States publicly attributed the cyberattack to North Korea, stating that "these disruptions put lives at risk."[97] Compromises of ITC supply chains can also pose a threat to organizations that rely on the tampered hardware as was alleged, for example, with some Supermicro microchips used in ITC manufacturing in China.[98]

Companies that rely on cloud services to store or transmit data may choose to use enhanced encryption to protect the communication and privacy, both internally and of their end customers. This, in turn, may impede law enforcement investigations if they are unable to access the encrypted data.[99] However, restrictions on the ability for a firm to use encryption may make a company vulnerable to cyberattacks or cybertheft, demonstrating the need for policies and regulations to balance competing objectives.

U.S. Digital Trade with Key Trading Partners

The European Union (EU) and China are large U.S. digital trade partners and each has presented various challenges for U.S. companies, consumers, and policymakers.

---

[96] Nick Kostov, Jeannette Neumeann, and Stu Woo, "Cyberattack Victims Begin to Assess Financial Damage," *Wall Street Journal*, May 14, 2017.

[97] Thomas P. Bossert, Assistant to the President for Homeland Security and Counterterrorism, "It's Official: North Korea Is Behind WannaCry," *Wall Street Journal*, December 18, 2017.

[98] Jordan Robertson and Michael Riley, "The Big Hack: How China Used a Tiny Chip to Infiltrate U.S. Companies," Bloomberg, October 4, 2018.

[99] For more information on encryption, see CRS Report R44187, *Encryption and Evolving Technology: Implications for U.S. Law Enforcement Investigations*, by Kristin Finklea, and CRS Report R44407, *Encryption: Selected Legal Issues*, by Richard M. Thompson II and Chris Jaikaran.

European Union

Differences in U.S. and EU policies have ramifications on digital flows and international trade. The two partners' varying approaches to digital trade, privacy, and national security, have, at times, threatened to disrupt U.S.-EU data flows.

The transatlantic economy is the largest in the world, and cross-border data flows between the United States and EU are the highest in the world. In between 2003 and 2017, total U.S.-EU trade in goods and services (exports plus imports) nearly doubled from $594 billion to $1.2 trillion.[100] ICT and potentially ICT-enabled services accounted for approximately $190 billion of U.S. exports to the EU in 2017.[101] The two sides also account for a significant portion of each other's e-commerce trade (see Figure 4).

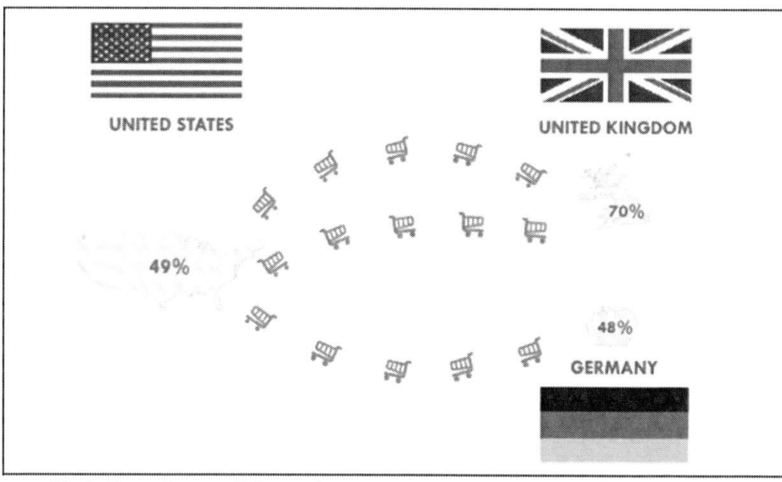

Source: Kati Souminen, "Where the Money Is: The Transatlantic Digital Market," CSIS, October 12, 2017.

Notes: 48% of German and 70% of UK shoppers purchase from U.S. e-commerce sites. 49% of U.S. e-commerce purchases are from UK sites.

Figure 4. Select U.S.-EU Cross-Border E-Commerce Purchases.

---

[100] See CRS In Focus IF10930, *U.S.-EU Trade and Investment Ties: Magnitude and Scope*, by Shayerah Ilias Akhtar.

[101] https://apps.bea.gov/iTable/iTable.cfm?ReqID=62&step=1.

The United States and EU account for almost half of each other's digitally deliverable service exports (e.g., business, professional, and technical services) and many of these services are incorporated into exported goods as part of GVCs (see Figure 5 and Figure 6).[102] The UK alone accounted for 23% of U.S. digitally deliverable services exports.[103] Almost 40% of the data flows between the United States and EU are through business and research networks.[104]

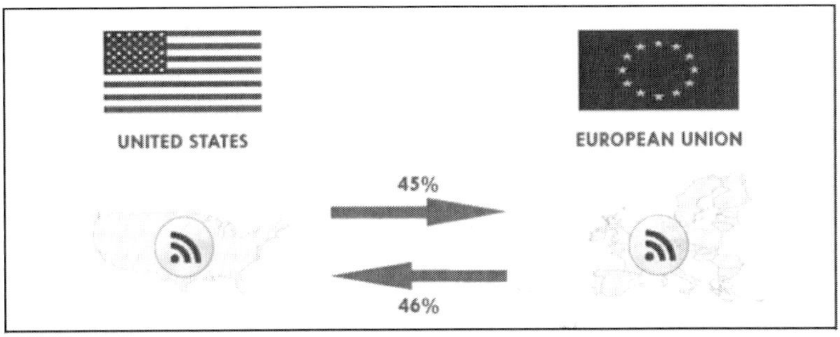

Source: "Where the Money Is: The Transatlantic Digital Market," CSIS, October 12, 2017.

Figure 5. Digitally Deliverable Service Exports 2017.

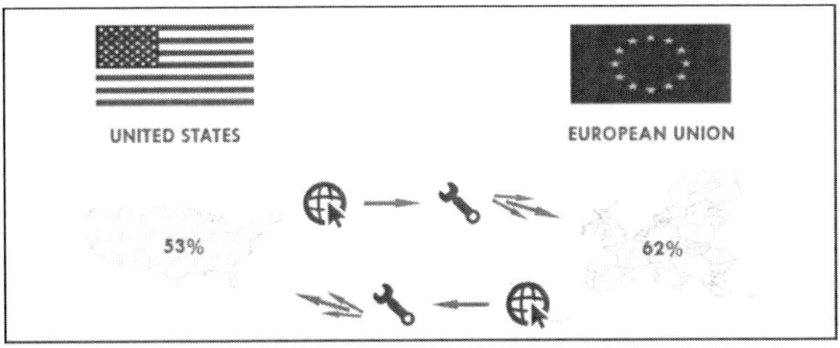

Source: "Where the Money Is: The Transatlantic Digital Market," CSIS, October 12, 2017.

Figure 6. Digitally Deliverable Services Incorporated into Global Value Chains.

[102] Where the Money Is: The Transatlantic Digital Market," CSIS, October 12, 2017.

[103] Ibid.

[104] All figures on U.S.-EU trade and data flows includes the United Kingdom (UK) as part of the EU. Without the UK, the statistics would be lower.

Despite close economic ties, differences between the United States and EU in their approaches to data flows and digital trade have caused friction in U.S.-EU economic and security relations. To address some of these differences, in 2013, the United States and the EU began, but did not conclude, negotiating a broad FTA. Negotiations included a number of digital trade issues such as market access for digital products, IPR protection and enforcement, cybersecurity, and regulatory cooperation, among other things.[105] On October 16, 2018, the Trump Administration notified Congress under Trade Promotion Authority (TPA) of its intent to enter into negotiations with the EU. The Administration's specific negotiating objectives envision a wide-ranging agreement, including addressing digital trade, along with trade in goods, services, agriculture, government procurement, and other rules, such as on IPR and investment.[106] However, no agreement exists on the scope of the negotiations. The EU negotiating mandates, in contrast, are narrower; they authorize EU negotiations with the United States to address industrial tariffs (excluding agricultural products) and nontariff regulatory barriers to make it easier for companies to prove that their products meet U.S. and EU technical requirements.[107]

The Administration also notified Congress under TPA of its intent to negotiate a trade agreement with the UK post-Brexit, and the corresponding specific negotiating objectives likewise envision a broad agreement addressing digital trade issues. The UK cannot formally negotiate or conclude a new agreement until it exits the EU, which has exclusive competence over trade policy and negotiates trade deals on

---

[105] Under the Obama Administration, a U.S. goal for T-TIP had been to develop "appropriate provisions to facilitate the use of electronic commerce to support goods and services trade, including through commitments not to impose customs duties on digital products or unjustifiably discriminate among products delivered electronically." USTR, "U.S. Objectives, U.S. Benefits in the Transatlantic Trade and Investment Partnership: A Detailed View," fact sheet, March 2014.

[106] Office of the U.S. Trade Representative, United States-European Union Negotiations: Summary of Specific Negotiating Objectives, January 2019. For more information, see CRS In Focus IF10931, *U.S.-EU Trade and Economic Issues*, by Shayerah Ilias Akhtar.

[107] Council of the EU, "Trade with the United States: Council authorises negotiations on elimination of tariffs for industrial goods and on conformity assessment," press release, April 15, 2019.

behalf of all EU member states. Details about the future UK-EU trade relationship remain largely unknown, and it is uncertain when and to what extent the UK will regain control of its national trade policy—a major objective for Brexit supporters. These factors directly shape prospects for a proposed bilateral U.S.-UK free trade agreement.[108]

EU-U.S. Privacy Shield

The United States and EU have different legal approaches to information privacy that extends into the digital world. After extensive negotiations, the EU-U.S. Privacy Shield entered into force on July 12, 2016, creating a framework to provide U.S. and EU companies a mechanism to comply with data protection requirements when transferring personal data between the EU and the United States.[109] Under the Privacy Shield program, U.S. companies can voluntarily self-certify compliance with requirements such as robust data processing obligations. The agreement includes obligations on the U.S. government to proactively monitor and enforce compliance by U.S. firms, establish an ombudsman in the U.S. State Department, and set specific safeguards and limitations on surveillance. The United States and Switzerland also agreed to the Swiss-U.S. Privacy Shield, which will be "comparable" to the EU-U.S. agreement.[110]

The Privacy Shield also involves an annual joint review by the United States and the EU, the second of which was completed in October 2018.[111] Under the review, the commission found that the Privacy Shield is working and that the United States had made improvements and changes since the first review. The Commission, however, also noted areas of concern and specific recommendations.

---

[108] CRS In Focus IF11123, *Brexit and Outlook for U.S.-UK Trade Agreement*, by Shayerah Ilias Akhtar.

[109] For more information on the Privacy Shield, see https://www.privacyshield.gov/Program-Overview.

[110] Lauren Cerulus, "Switzerland and U.S. strike 'privacy shield' data transfer deal," *Politico Pro*, January 11, 2017.

[111] European Commission, "Report from the Commission to the European Parliament and the Council," COM(2018) 860 final, December 19, 2018.

General Data Protection Regulation (GDPR)

The EU's General Data Protection Regulation (GDPR), effective May 2018, established rules for EU member states to safeguard individuals' personal data. The GDPR is a comprehensive privacy regime that builds on previous EU data protection rules. It grants new rights to individuals to control personal data and creates specific new data protection requirements. The GDPR applies to (1) all businesses and organizations with an EU establishment that process (perform operations on) personal data of individuals (or "data subjects") in the EU, regardless of where the actual processing of the data takes place; and (2) entities outside the EU that offer goods or services (for payment or for free) to individuals in the EU or monitor the behavior of individuals in the EU. These measures have raised concerns about the GDPR's extraterritorial implications.

While the GDPR is directly applicable at the EU member state level, individual countries are responsible for establishing some national-level rules and policies as well as enforcement authorities, and some are still in the process of doing so. As a result, some U.S. stakeholders have voiced concern about a lack of clarity and inadequate country compliance guidelines, as well as about the potential high cost of data storage and processing needed for compliance. Despite the lack of precise guidance, many companies have taken steps to implement its requirements. For example, Amazon touts its compliance with GDPR requirements and aims to assist its Amazon Web Services (AWS) corporate customers, many of whom are small and medium businesses, with their own compliance.[112] It can be more challenging for SMEs to fully understand GDPR and comply with its notification and other requirements such as an individual's "right to be forgotten" and on data portability; there are indications that some U.S. businesses have chosen to exit the EU market.[113]

Some experts contend that the GDPR may effectively set new global data privacy standards, since many companies and organizations are striving for GDPR compliance to avoid being shut out of the EU market,

---

[112] See https://aws.amazon.com/compliance/gdpr-center/.

[113] "Websites not available in the European Union after GDPR," VerifiedJoseph.com, July 11, 2018, updated November 16, 2018, https://data.verifiedjoseph.com/dataset/websites-notavailable-eu-gdpr.

fined, or otherwise penalized. In addition, some countries outside of Europe are imitating all or parts of the GDPR in their own privacy regulatory and legislative efforts. European Data Protection Authorities may have reinforced U.S. companies' concerns by initiating several enforcement actions in the fall of 2018, including a €50 million (approximately $57 million) fine on Google.[114]

Digital Single Market (DSM)

Like the GDPR, EU policymakers are attempting to bring more harmonization across the region through the Digital Single Market (DSM). The DSM is an ongoing effort to unify the EU market, facilitate trade, and drive economic growth. The DSM's three pillars revolve around better online access to cross-border digital goods and services; a regulatory environment supporting investment and fair competition; and driving growth through investment in infrastructure, human capital, research, and innovation. Among its initiatives is a mandate to allow cross-border flows for nonpersonal data within the EU (with limited exceptions), but not necessarily externally.

China

China presents a number of significant opportunities and challenges for the United States in digital trade. The modernization of the Chinese economy, coupled with a large and increasingly prosperous population, has led to a surge in the number of Chinese Internet users and made China a major source of global ecommerce. China's internet users grew from 21.5 million in 2000 to 829 million as of March 2019, and this trend will likely continue, given China's relatively low internet penetration rate (see Figure 7).[115] China's online retail sales in 2018 totaled $1.1 trillion (more

---

[114] For more information on GDPR, see CRS In Focus IF10896, *EU Data Protection Rules and U.S. Implications*, by Rachel F. Fefer and Kristin Archick.

[115] *Internet World Statistics* at https://www.internetworldstats.com/top20.htm.

than double the U.S. level at $505 billion) and were the world's largest.[116] E-Marketer predicts that China's e-commerce retail sales will reach $1.99 trillion in 2019, accounting for 35.3% of total sales and 55.8% of global online sales.[117]

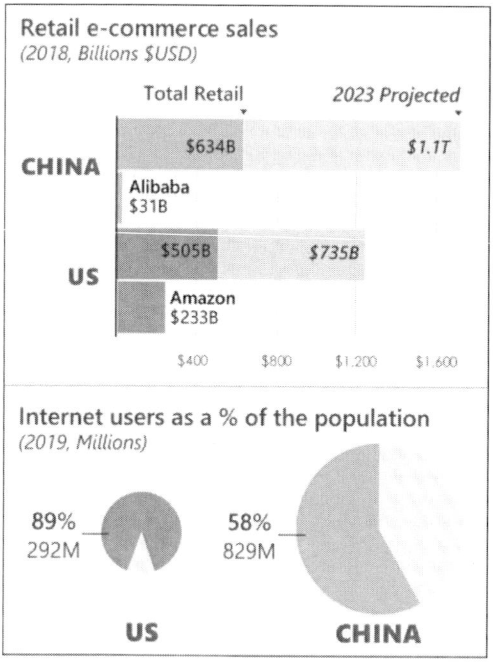

Source: U.N. population statistics, Statista.com, Internetworldstats.com.

Figure 7. The U.S. and China Digital Trade Markets.

U.S. firms may benefit from expanding digital trade in China, but they may also face numerous challenges in the Chinese market. The USTR's 2019 report on foreign trade barriers included a digital trade fact sheet that cited countries and practices of "key concern."[118] Three Chinese digital policies were listed, including its restrictions on cross-border data flows

[116] Statista.com.

[117] E-Marketer, Newsroom, *2019: China to Surpass US in Total Retail Sales*, January 23, 2019, available at https://www.emarketer.com/newsroom/index.php/2019-china-to-surpass-us-in-total-retail-sales/.

[118] USTR, *Fact Sheet on 2019 National Trade Estimate: Key Barriers to Digital Trade*, March 2019, available at https://ustr.gov/about-us/policy-offices/press-office/fact-sheets/2019/march/fact-sheet-2019-national-trade-estimate.

and data localization requirements; extensive web filtering and blocking of legitimate sites, including blocks 10 of the top 30 global sites and up to 10,000 sites in total, affecting billions of dollars in potential U.S. business; and cloud computing restrictions and requirements to partner with a Chinese firm to enter the market and to transfer technology and IP to the partner.[119]

The American Chamber of Commerce in China (AmCham China) 2019 business survey found that 73% of respondents who were engaged in technology and R&D-intensive industries stated that they faced significant or somewhat significant market barriers in China. The lack of sufficient IPR protection (cited by 35% of respondents) and restrictive cybersecurity-related policies (cited by 27% of respondents) ranked among the top three factors prohibiting firms from increasing innovation activities in China. The survey reflected significant concerns by member firms over eight Chinese ICT policies and restrictions (such as internet restrictions and censorship, IPR theft, and data localization requirements), with 72% to 88% of respondents stating that such measures impacted their competiveness and operations in China either somewhat or severely (see Table 1).

A Digital Trade Restrictiveness Index (DTRI) of 65 economies created by the European Centre for International Political Economy found China to have the most restrictive digital policies, followed by Russia, India, Indonesia, and Vietnam.[120] The index report noted:

> China applies the most restrictive digital trade measures in many areas, including public procurement, foreign investment, Intellectual Property Rights (IPRs), competition policy, intermediary liability, content access and standards. The restrictions do not only impose higher costs for trading digital goods and services, they can also block digital trade altogether in certain sectors. In addition, China's data policies are

---

[119] Examples of blocked sites include Google services (e.g., Gmail and YouTube), Twitter, Instagram, Facebook, the Washington Post, the New York Times, and the Wall Street Journal.

[120] The index was developed based on four factors: fiscal restrictions and market access, establishment restrictions, data restrictions, and trading restrictions. See http://globalgovernanceprogramme.eui.eu/wp-content/uploads/2018/09/DTRI-final.pdf.

extremely burdensome for companies, and the country also applies some quantitative trade restrictions and restrictions on e-commerce.[121]

Table 1. AmCham China Business Survey: Percent of Respondents who said Certain Chinese IT Policies affected their Operations and Competitiveness in China somewhat or severely

IT-related issues and practices	% of respondents
Slow cross border internet speed	88
Restricted access to online tools such as software	86
Cross-border internet access by virtual private networks (VPN)	83
Data security/IP leakage	79
Cybersecurity rules protecting critical information infrastructure/important data	75
Data privacy regulations	75
Internet censorship and restrictions on information publishing/sharing	73
Data localization requirements	72

Source: 2019 AmCham China Business Survey.

Internet Governance and the Concept of "Internet Sovereignty"

The Chinese government has sought to advance its views on how the internet should be expanded to promote trade, but also to set guidelines and standards over the rights of governments to regulate and control the internet, a concept it has termed "Internet Sovereignty."[122] The Chinese government appears to have first advanced a policy of "Internet Sovereignty" around June 2010 when it issued a White Paper titled "the Internet of China," which stated the following:

> Within Chinese territory the Internet is under the jurisdiction of Chinese sovereignty. The Internet sovereignty of China should be respected and protected. Citizens of the People's Republic of China and foreign citizens, legal persons and other organizations within Chinese territory have the right and freedom to use the Internet; at the same time,

[121] The top five most open economies were New Zealand, Iceland, Norway, Ireland, and Hong Kong.

[122] Originally, China appeared to be mainly focused on establishing the rules of the road for the Internet in China, but over the past few years it appears to be advancing its vision of Internet sovereignty globally.

> they must obey the laws and regulations of China and conscientiously protect Internet security.[123]

In 2014, the Chinese government established the Central Internet Security and "Informatization" Leading Group, headed by Chinese president Xi Jinping, to "strengthen China's Internet security and build a strong cyberpower." A year later, President Xi addressed an internet conference, stating "we should respect the right of individual countries to independently choose their own path of cyber development, model of cyber regulation and Internet public policies, and participate in international cyberspace governance on an equal footing."[124]

Some analysts contend that China's internet sovereignty initiative represents an assertion that the government has the right to fully control the internet within China. Some see this as an attempt by the government to control information that is deemed a threat to social stability, in violation of the right to freedom of speech, which is guaranteed in China's Constitution. Other critics of China's internet sovereignty policy view it as an attempt by the government to limit market access by foreign internet, digital, and high technology firms in China, in order to boost Chinese firms and reduce China's dependence on foreign technology.

Cyber-Theft of U.S. Trade Secrets

China is considered by most analysts to be the largest source of global theft of IP and a major source of cybertheft of U.S. trade secrets, including by government entities. To illustrate, a 2011 report by the U.S. Office of the Director of National Intelligence (DNI) stated: "Chinese actors are the world's most active and persistent perpetrators of economic espionage. U.S. private sector firms and cybersecurity specialists have reported an onslaught of computer network intrusions that have originated in China,

---

[123] The People's Daily, *Full Text: The Internet in China*, June 8, 2010, available at http://en.people.cn/90001/90776/ 90785/7017202.html.

[124] Ministry of Foreign Affairs of the People's Republic of China, *Remarks by H.E. Xi Jinping President of the People's Republic of China At the Opening Ceremony of the Second World Internet Conference*, December 16, 2015, available at http://www.fmprc.gov.cn/mfa_eng/wjdt_665385/zyjh_665391/t1327570.shtml.

but the IC (Intelligence Community) cannot confirm who was responsible." The report goes on to warn that

> China will continue to be driven by its longstanding policy of "catching up fast and surpassing" Western powers. The growing interrelationships between Chinese and U.S. companies—such as the employment of Chinese-national technical experts at U.S. facilities and the off-shoring of U.S. production and R&D to facilities in China—will offer Chinese government agencies and businesses increasing opportunities to collect sensitive US economic information.[125]

In May 2014, the U.S. Department of Justice issued a 31-count indictment against five members of the People's Liberation Army for cyber-espionage and other offenses that allegedly targeted five U.S. firms and a labor union for commercial advantage, the first time the Federal government had initiated such action against state actors.[126]

In April 2015, President Obama issued Executive Order 13964 authorizing certain sanctions against "persons engaging in significant malicious cyber-enabled activates."[127] This led to China sending a high-level delegation to Washington, DC, and, on September 25, 2015, Presidents Obama and Xi announced that they had reached an agreement on cyber-security and trade secrets that stated that neither country's government "will conduct or knowingly support cyber-enabled theft of IP, including trade secrets or other confidential business information, with the intent of providing competitive advantages to companies or commercial sectors."[128] Specifically, the two sides agreed to

---

[125] DNI, Office of the National Counterintelligence Executive, *Foreign Spies Stealing U.S. Economic Secrets in Cyberspace, Report to Congress on Foreign Economic Collection and Industrial Espionage*: 2009-2011, October 2011.

[126] U.S. Department of Justice, at http://www.justice.gov/iso/opa/resources/5122014519 132358461949.pdf.

[127] A copy can be found at http://www.treasury.gov/resource-center/sanctions/Programs/ Documents/cyber_eo.pdf. The EO was extended for an additional year by President Obama on March 29, 2016.

[128] The November 2015 meeting of the G-20 countries (which includes China) included this language in its communique: "In the ICT environment, just as elsewhere, states have a special responsibility to promote security, stability, and economic ties with other nations. In support of that objective, we affirm that no country should conduct or support ICT-enabled theft of intellectual property, including trade secrets or other confidential business

- Not conduct or knowingly support cyber-enabled theft of IP, including trade secrets or other confidential business information, with the intent of providing competitive advantages to companies or commercial sectors;
- Establish a high-level joint dialogue mechanism on fighting cybercrime and related issues;
- Work together to identify and promote appropriate norms of state behavior in cyberspace internationally; and
- Provide timely responses to requests for information and assistance concerning malicious cyber activities.[129]

The two sides also agreed to set up a high-level dialogue mechanism (which would take place twice a year) to address cybercrime and improve two-way communication when cyber-related concerns arise (including the creation of a hotline). The first meeting of the *U.S.-China High-Level Joint Dialogue on Cybercrime and Related Issues* was held in December 2015. China and the United States reached agreement on a document establishing guidelines for requesting assistance on cybercrime or other malicious cyber activities and for responding to such requests. Two more meetings were held in 2016. The dialogue was continued in October 2017 under the Trump Administration.[130] The Administration's Section 301 trade dispute between the United States and China may have led to a suspension of the dialogue (see below).[131]

It is difficult to assess the effectiveness of the September 2015 U.S.-China cyber agreement in reducing the level of Chinese cyber intrusions against U.S. entities seeking to steal trade secrets as no official U.S. statistics on such activities are publicly available. In August 2018, the U.S.

information, with the intent of providing competitive advantages to companies or commercial sectors."

[129] The White House, Fact Sheet, President Xi Jinping's State Visit to the United States, September 25, 2015, available at https://obamawhitehouse.archives.gov/the-press-office/2015/09/25/fact-sheet-president-xi-jinpings-state-visit-unitedstates.

[130] See U.S. Department of Justice, *Press Release*, October 6, 2017, at https://www.justice.gov/opa/pr/first-us-chinalaw-enforcement-and-cybersecurity-dialogue.

[131] The Diplomat, *Another US-China Dialogue Bites the Dust*, October 2, 2018, at https://thediplomat.com/2018/10/ another-us-china-dialogue-bites-the-dust/.

Deputy Director of the Cyber Threat Intelligence Integration Center stated that "the intelligence community and private-sector security experts continue to identify ongoing cyber activity from China, although at volumes significantly lower than before the bilateral U.S.-China cyber commitments of September 2015."[132] In October 2018, CrowdStrike, a U.S. cybersecurity technology company, identified China as "the most prolific nation-state threat actor during the first half of 2018."[133] It found that Chinese entities had made targeted intrusion attempts against multiple sectors of the economy. In December 2018, U.S. Assistant Attorney General John C. Demers stated at a Senate hearing that from 2011-2018, China was linked to more than 90% of the Justice Department's cases involving economic espionage and two-thirds of its trade secrets cases.[134]

Cybersecurity Laws

According to the USTR's 2017 report on China's WTO accession, China has not fulfilled all of its WTO market opening commitments. The USTR cited "significant declines in commercial sales of foreign ICT products and services in China," as evidence that China continued to maintain "mercantilist policies under the guise of cybersecurity."[135]

The Chinese government pledged not to use recently enacted cyber and national security laws and regulations to unfairly burden foreign ICT firms, or to discriminate against foreign ICT firms in the implementation of various policy initiatives to promote indigenous innovation in China. Some Chinese laws or proposals include language stating that critical information

---

[132] Office of the Director of National Intelligence, *Statement for the Record Mr. Michael Moss, Deputy Director Cyber Threat Intelligence Integration Center on "Cyber Threats to Our Nation's Critical Infrastructure,"* August 21, 2018, available at https://www.dni.gov/index.php/ctiic-newsroom/item/1899-statement-for-the-record-mr-michael-moss-for-confirmation-before-the-senate-select-committee-on-crime-and-terrorism-to-be-deputy-director-of-the-cyber-threat-intelligence-integration-center.

[133] CrowdStrike, *CrowdStrike Report Reveals Cyber Intrusion Trends from Elite Team of Threat Hunters*, October 9, 2019, at https://www.crowdstrike.com/resources/news/crowdstrike-report-reveals-cyber-intrusion-trends-from-eliteteam-of-threat-hunters/.

[134] U.S. Department of Justice, Statement of John C. Demers, Assistant Attorney General, National Security Division, U.S. Department of Justice Before the Committee on the Judiciary, United States Senate, December 12, 2018, at https://www.judiciary.senate.gov/imo/media/doc/12-12-18%20Demers%20Testimony.pdf.

[135] USTR, *2017 Report to Congress on China's WTO Compliance*, January 2018, p. 3.

infrastructure should be "secure and controllable," an ambiguous term that has not been precisely defined by Chinese authorities. Other proposals of concern to U.S. firms appear to lay out policies that would require foreign ICT firms to hand over proprietary information.

Examples of measures of concern to foreign ICT firms include

- Cybersecurity Law, passed by the government on November 7, 2016 (effective June 1, 2017), ascertains the principles of cyberspace sovereignty;[136] defines the security-related obligations of network product and service providers; further enhances the rules for protection of personal information; establishes a framework of security protection for "critical information infrastructure"; and establishes regulations pertaining to cross-border transmissions of important data by critical information infrastructure.[137]

Some analysts have expressed concerns that one of the main goals of the new law is to promote the development of indigenous technologies and impose restrictions on foreign firms, and many multinational companies continue to voice concerns about the lack of clarity of the law's requirements, how the law will be interpreted and implemented through subsequent regulations, and to what extent it will impact their operations in China.

- National Security Law, enacted in July 2015, emphasizes the state's role in driving innovation and reviewing "foreign commercial investment, special items and technologies, internet information technology products and services, projects involving

---

[136] Article 1 states: "This law is formulated so as to ensure network security, to safeguard cyberspace sovereignty, national security and the societal public interest, to protect the lawful rights and interests of citizens, legal persons and other organizations, and to promote the healthy development of economic and social informatization."

[137] Deloitte, "A new era for Cybersecurity in China," November 2017, available at https://www2.deloitte.com/cn/en/ pages/risk/articles/new-era-cybersecurity-law.html.

national security matters, as well as other major matters and activities, that impact or might impact national security."[138]

Such restrictions could have a significant impact on U.S. ICT firms. According to BEA, U.S. exports of ICT services and potentially ICT-enabled services (i.e., services that are delivered remotely over ICT networks) to China totaled $18.7 billion in 2017.[139]

Section 301 Action against China over Intellectual Property and Innovation Issues

Concerns over China's policies on IP, technology, and innovation policies led the Trump Administration, in August 2017, to launch a Section 301 investigation of those policies.[140] On March 22, 2018, President Trump signed a Memorandum on Actions by the United States Related to the Section 301 Investigation that identified four broad IPR-related policies that justified U.S. action under Section 301, stating that China

1. Uses joint venture requirements, foreign investment restrictions, and administrative review and licensing processes to force or pressure technology transfers from American companies;
2. Uses discriminatory licensing processes to transfer technologies from U.S. companies to Chinese companies;

---

[138] Article 59, translation from the Council on Foreign Relations, *National Security Law of the People's Republic of China*, July 1, 2015, http://www.cfr.org/homeland-security/national-security-law-peoples-republic-china/p36775.

[139] See, BEA, *International Trade Data, U.S. Trade in Services*, https://apps.bea.gov/iTable/iTable.cfm?ReqID=62& step=1#reqid=62&step=1&isuri=1&6210=4.

[140] Sections 301 through 310 of the Trade Act of 1974, as amended, commonly referred to as "Section 301," procedures apply to foreign acts, policies, and practices that the USTR determines either (1) violates, or is inconsistent with, a trade agreement; or (2) is unjustifiable and burdens or restricts U.S. commerce, and sets procedures and timetables for actions based on the type of trade barrier(s) addressed.

3. Directs and facilitates investments and acquisitions which generate large-scale technology transfer; and
4. Conducts and supports cyberintrusions into U.S. computer networks to gain access to valuable business information.

The USTR estimates such policies cost the U.S. economy at least $50 billion annually. Under the Section 301 action, the Administration proposed to (1) implement 25% ad valorem tariffs on certain Chinese imports (which in sum are comparable to U.S. trade losses); (2) initiate a WTO dispute settlement case against China's "discriminatory" technology licensing (which it did on March 23, 2018); and (3) propose new investment restrictions on Chinese efforts to acquire sensitive U.S. technology.[141] The Administration did not act on the last issue after Congress passed the Foreign Investment Risk Review Modernization Act of 2018 (FIRRMA) (P.L. 115- 232) in August 2018 to modernize the existing U.S. review process of foreign investments in terms of national security. Among its changes, FIRRMA expanded the types of investment subject to review, including certain noncontrolling investments in "critical technology."[142]

The Trump Administration subsequently imposed tariff hikes on $250 billion worth of imports from China in three separate stages in 2018, while China increased tariffs on $110 billion worth of imports from the United States (See Figure 8).[143] In May 2019, the United States increased the tariff levels on the third tranche of products imported from China. China subsequently increased its tariff levels on its third tranche.

---

[141] For more information on the Section 301 investigation, see CRS In Focus IF10708, *Enforcing U.S. Trade Laws: Section 301 and China*, by Wayne M. Morrison.

[142] For more information on FIRMMA, see CRS In Focus IF10952, *CFIUS Reform: Foreign Investment National Security Reviews*, by James K. Jackson and Cathleen D. Cimino-Isaacs.

[143] These import hikes were as follows: 25% on $34 billion (July 6); 25% on $16 billion (August 23), and 10% on $200 billion (September 24). China's first two stages of retaliation matched U.S. levels in import and tariff levels while its third stage tariff hikes ranged from 5% to 10% on $60 billion worth of imports.

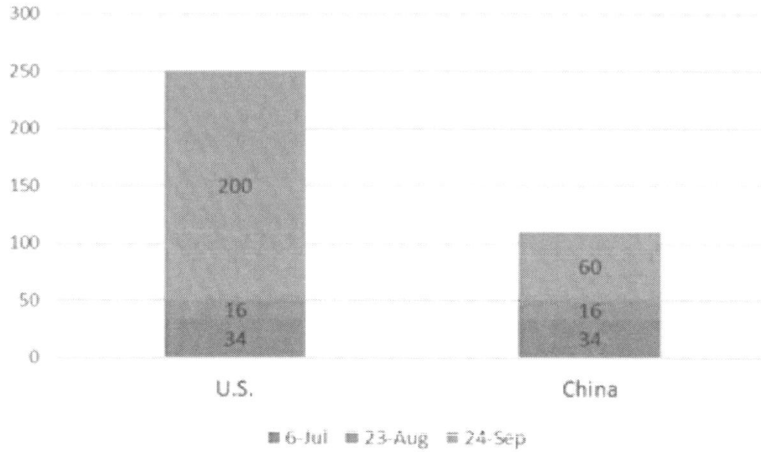

Source: USTR and Chinese Ministry of Commerce. Notes: Tariff rates vary.

Figure 8. Three Rounds of U.S.-China Tariff Hikes in 2018. Estimated Value of Goods Impacted ($billions) and effective dates.

DIGITAL TRADE PROVISIONS IN TRADE AGREEMENTS

As the above analysis of EU and China policies demonstrates, there is not a single set of international rules or disciplines that govern key digital trade issues, and the topic is treated inconsistently, if at all, in trade agreements. As digital trade has emerged as an important component of trade flows, it has risen in significance on the U.S. trade policy agenda and that of other countries.

Given the stalemate in comprehensive WTO multilateral negotiations, trade agreements have not kept pace with the complexities of the digital economy and digital trade is treated unevenly in existing WTO agreements. More recent bilateral and plurilateral deals have started to address digital trade policies and barriers more comprehensively. The use of digital trade provisions in bilateral and plurilateral trade negotiations may help spur interest in the creation of future WTO frameworks that focus on digital trade and provide input for ongoing plurilateral negotiations occurring in the aegis of the WTO (see below).

WTO Provisions

While no comprehensive agreement on digital trade exists in the WTO, other WTO agreements cover some aspects of digital trade and new plurilateral negotiations may set new rules and disciplines.

General Agreement on Trade in Services (GATS)

The WTO General Agreement on Trade in Services (GATS) entered into force in January 1995, predating the current reach of the internet and the explosive growth of global data flows. GATS includes obligations on nondiscrimination and transparency that cover all service sectors. The market access obligations under GATS, however, are on a "positive list" basis in which each party must specifically opt in for a given service sector to be covered.[144]

As GATS does not distinguish between means of delivery, trade in services via electronic means is covered under GATS. While GATS contains explicit commitments for telecommunications and financial services that underlie e-commerce, digital trade and information flows and other trade barriers are not specifically included. Given the positive list approach of GATS, coverage across members varies and many newer digital products and services did not exist when the agreements were negotiated. To address advances in technology and services, the Committee on Specific Commitments is examining how certain new online services, such as platform services, or specific regulations, such as data localization, could be classified and scheduled within GATS.[145]

Declaration on Global Electronic Commerce

In May 1998, WTO members established the "comprehensive" Work Programme on Electronic Commerce and established a temporary customs duties moratorium on electronic transmission that has been extended

---

[144] For more information, see https://www.wto.org/english/tratop_e/serv_e/serv_e.htm, and CRS Report R43291, *U.S. Trade in Services: Trends and Policy Issues*, by Rachel F. Fefer.

[145] World Trade Organization, "WTO members hold latest "cluster" of services meetings," March 21, 2019.

multiple times.[146] While multiple members submitted proposals to advance multilateral digital trade negotiations under the Work Programme, no clear path forward was identified.

Information Technology Agreement (ITA)

The WTO Information Technology Agreement (ITA) aims to eliminate tariffs on the goods that power and utilize the internet, lowering the costs for companies to access technology at all points along the value chain. Originally concluded in 1996, the ITA was expanded to further cut tariffs beginning in July 2016. The expanded ITA is a plurilateral agreement among 54 developed and developing WTO members who account for over 90% of global trade in these goods. Some WTO members, such as Vietnam and India, are party to the original ITA, but did not join the expanded agreement. Like the original ITA, the benefits of the expanded agreement will be extended on a most-favored nation (MFN) basis to all WTO members.

Under the expanded ITA, the parties agreed to review the agreement's scope in the future to determine if additional product coverage is warranted as technology evolves. While the WTO ITA has expanded trade in the technology products that underlie digital trade, it does not tackle the nontariff barriers that can pose significant limitations.

Agreement on Trade-Related Aspects of Intellectual Property Rights (TRIPS)

The TRIPS Agreement, in effect since January 1, 1995, provides minimum standards of IPR protection and enforcement. The TRIPS Agreement does not specifically cover IPR protection and enforcement in the digital environment, but arguably has application to the digital environment and sets a foundation for IPR provisions in subsequent U.S. trade negotiations and agreements, many of which are "TRIPS-plus."

The TRIPS Agreement covers copyrights and related rights (i.e., for performers, producers of sound recordings, and broadcasting

---

[146] For more information, see https://www.wto.org/english/tratop_e/ecom_e/ecom_briefnote_e.htm.

organizations), trademarks, patents, trade secrets (as part of the category of "undisclosed information"), and other forms of IP. It builds on international IPR treaties, dating to the 1800s, administered by the World Intellectual Property Organization, or WIPO (see below). TRIPS incorporates the main substantive provisions of WIPO conventions by reference, making them obligations under TRIPS. WTO members were required to fully implement TRIPS by 1996, with exceptions for developing country members by 2000 and least-developed-country (LDC) members until July 1, 2021, for full implementation.[147]

TRIPS aims to balance rights and obligations between protecting private rights holders' interests and securing broader public benefits. Among its provisions, the TRIPS section on copyright and related rights includes specific provisions on computer programs and compilations of data. It requires protections for computer programs—whether in source or object code—as literary works under the WIPO Berne Convention for the Protection of Literary and Artistic Works (Berne Convention). TRIPS also clarifies that databases and other compilations of data or other material, whether in machine readable form or not, are eligible for copyright protection even when the databases include data not under copyright protection.[148]

Like the GATS, TRIPS predates the era of ubiquitous internet access and commercially significant e-commerce. TRIPS includes a provision for WTO members to "undertake reviews in the light of any relevant new developments which might warrant modification or amendment" of the agreement. The TRIPS Council has engaged in discussions on the agreement's relationship to electronic commerce as part of the WTO Work Programme on Electronic Commerce, focusing on protection and enforcement of copyright and related rights, trademarks, and new

---

[147] For pharmaceutical products, the implementation period has been extended until January 1, 2033.

[148] WTO, "Overview: The TRIPS Agreement," https://www.wto.org/english/tratop_e/trips_e/intel2_e.htm. For more information, see CRS Report RL34292, *Intellectual Property Rights and International Trade*, by Shayerah Ilias Akhtar and Ian F. Fergusson.

technologies and access to these technologies; new activity by the TRIPS Council to this end appears to be limited in recent years.[149]

World Intellectual Property Organization (WIPO) Internet Treaties

The World Intellectual Property Organization (WIPO) has been a primary forum to address IP issues brought on by the digital environment since the TRIPS Agreement. The WIPO Copyright Treaty and WIPO Performances and Phonograms Treaty—often referred to jointly as the WIPO "Internet Treaties"—established international norms regarding IPR protection in the digital environment. These treaties were agreed to in 1996 and entered into force in 2002, but are not enforceable, including under WTO dispute settlement. Shaped by TRIPS, the WIPO Internet Treaties are intended to clarify that existing rights continue to apply in the digital environment, to create new online rights, and to maintain a fair balance between the owners of rights and the general public.[150]

Key features of the WIPO Internet Treaties include provisions for legal protection and remedies against circumventing TPMs, such as encryption, and against the removal or alteration of rights management information (RMI), which is data identifying works or their authors necessary for them to manage their rights (e.g., for licenses and royalties). The liability of online service providers and other communication entities that provide access to the internet was contested in the negotiations on the WIPO Internet Treaties. In the end, WIPO Internet Treaties leave it to the discretion of national governments to develop the legal parameters for ISP liability.[151]

As of March 2019, the WIPO Internet Treaties had 96 contracting parties. The United States implemented the WIPO Internet Treaties

---

[149] WTO, General Council, "Item 6—Work Programme on Electronic Commerce—Review of Progress," WT/GC/W/701, July 24, 2015; and WTO, General Council, "Item 4—Work Programme on Electronic Commerce— Review of Progress," WT/GC/W/756, December 17, 2018.

[150] BSA, *Powering the Digital Economy: A Trade Agenda to Drive Growth*; and BSA, *Shadow Market: 2011 BSA Global Software Piracy Study*, May 2012.

[151] U.S. Congress, Senate Committee on Foreign Relations, *WIPO Copyright Treaty (WCT) (1996) and WIPO Performances and Phonograms Treaty (1996)*, Report to accompany treaty document 105-17, 105th Cong., 2nd sess., October 14, 1998, S.Exec. Rept. 105-25.

through the Digital Millennium Copyright Act of 1998 (DMCA) (H.R. 2281), which set new standards for protecting copyrights in the digital environment, including prohibiting the circumvention of antipiracy measures incorporated into copyrighted works and enforcing such violations through civil, administrative, and criminal remedies.[152] The DMCA also, among other things, limits remedies available against ISPs that unknowingly transmit copyright infringing information over their networks by creating certain "safe harbors."[153] India was one of the latest countries to join the treaties, entering them into force on December 25, 2018. The United States continues to call on trading partners, such as Turkey and Mexico, to fully implement the WIPO Internet Treaties.[154]

WTO Plurilateral Effort

On the sidelines of the WTO Ministerial Conference, in December 2017, the United States, as part of a group of over 70 WTO members, agreed to "initiate exploratory work together toward future WTO negotiations on trade related aspects of electronic commerce."[155] The U.S. objectives include market access, data flows, nondiscriminatory treatment of digital products, protection of intellectual property and digital security measures, and intermediary liability, among others.[156]

The group formally launched the e-commerce initiative in January 2019.[157] The official joint statement lists includes advanced economies such as the United States, the EU, and Australia, and also several developing countries such as China and Brazil. India stated it will not join, preferring to maintain its flexibility to favor domestic firms, limit foreign

---

[152] See P.L. 105-304.

[153] For more information on this statute, see CRS Report R43436, *Safe Harbor for Online Service Providers Under Section 512(c) of the Digital Millennium Copyright Act*, by Brian T. Yeh.

[154] USTR, *2017 Special 301 Report*, April 2017.

[155] WTO, "Joint Statement on Electronic Commerce," December 13, 2017, https://ustr.gov/sites/default/files/files/Press/Releases/Joint%20Statement%20on%20Electronic%20Commerce.pdf.

[156] The United States, "Joint Statement on Electronic Commerce Initiative," WTO, April 12, 2018.

[157] WTO Joint Statement on Electronic Commerce, WT/L/1056, January 25, 2019.

market access, and raise revenue in the future through potential customs duties.[158]

After the meeting, the U.S. Trade Representative's (USTR) statement emphasized the need for a high-standard agreement that includes enforceable obligations.[159] The EU noted e-signatures, customs duties, forced disclosure of source code, and data localization measures among the potential new rules to be discussed.[160] Some analysts raise concerns that the EU may seek more limited commitments on issues such as cross-border data flows. China has proposed the negotiations be limited to exploratory discussions rather than establishing obligations on topics such as data flows and data storage.[161] The negotiating parties continue to discuss the scope of any potential agreement, but the outlook may be challenging given the different approaches and policies especially among the U.S., EU, and China.

U.S. Bilateral and Plurilateral Agreements

As traditional trade policy does not clearly reflect the pervasiveness of the digital economy, and data is increasingly incorporated into international trade, the line between goods and services, and the application of the existing multilateral trade agreement system, is not always clear. As discussed above, the WTO agreements provide limited treatment of some aspects of digital trade. The United States has sought to establish new rules and disciplines on digital trade in its bilateral and plurilateral trade negotiations.

---

158 Subhayan Chakraborty, "India refuses to join e-commerce talks at WTO, says rules to hurt country," *The Business Standard*, February 25, 2019.

159 USTR, "USTR Robert Lighthizer on the Joint Statement on Electronic Commerce," January 25, 2019, https://ustr.gov/about-us/policy-offices/press-office/press-releases/2019/january/ustr-robert-lighthizer-joint.

160 European Commission, "75 countries launch WTO talks on e-commerce," Press Release Database, January 25, 2019.

161 WTO Joint Statement on Electronic Commerce, INF/ECOM/19, April 23, 2019.

Existing U.S. Free Trade Agreements (FTAs)

The United States has included an e-commerce chapter in its FTAs since it signed an agreement with Singapore in 2003 that has progressively evolved.[162] The e-commerce chapter of U.S. FTAs usually begins by recognizing e-commerce as an economic driver and the importance of removing trade barriers to e-commerce.[163] Most chapters contain provisions on nondiscrimination of digital products, prohibition of customs duties, transparency, and cooperation topics such as SMEs, cross-border information flows, and promoting dialogues to develop e-commerce. Some of the FTAs also include cooperation on consumer protection, as well as providing for electronic authentication and paperless trading. All FTAs allow certain exceptions to ensure that each party is able to achieve legitimate public policy objectives, protecting regulatory flexibility.

Electronic Commerce Chapter Article 1 in U.S. FTAs

"The Parties recognize the economic growth and opportunity that electronic commerce provides, the importance of avoiding barriers to its use and development, and the applicability of the WTO Agreement to measures affecting electronic commerce."

The U.S.-South Korea FTA (KORUS) contains the most robust digital trade provisions in a U.S. FTA currently in force.[164] In addition to the provisions in prior FTAs, KORUS includes provisions on access and use of the internet to ensure consumer choice and market competition. Most significantly, KORUS was the first attempt in a U.S. FTA to explicitly address cross-border information flows. The e-commerce chapter contains an article that recognizes its importance and discourages the use of barriers to cross-border data but does not explicitly mention localization requirements. The financial services chapter of KORUS also contains a

162 https://ustr.gov/sites/default/files/uploads/agreements/fta/singapore/asset_upload_file708_4036.pdf.

163 This statement was used in U.S. free trade agreements with Australia, Bahrain, Colombia, Central America and the Dominican Republic, Morocco, Oman, Panama, Peru, and South Korea. Chile used a slightly different text.

164 For more information on KORUS, see CRS Report RL34330, *The U.S.-South Korea Free Trade Agreement (KORUS FTA): Provisions and Implementation*, coordinated by Brock R. Williams.

specific, enforceable commitment to allow cross-border data flows "for data processing where such processing is required in the institution's ordinary course of business."[165]

In 2018, the Trump Administration and South Korea agreed to limited modifications of the agreement, but no changes were made to provisions directly impacting digital trade.

United States-Mexico-Canada Agreement (USMCA)

The released text of the proposed USMCA with Canada and Mexico aims to revise and update the trilateral North American Free Trade Agreement (NAFTA), and illustrates the Trump Administration's approach to digital trade.[166] The final text of the agreement pulls from and builds on many of the provisions from the Trans-Pacific Partnership (TPP) negotiated under President Obama which the United States did not ratify.[167] The provisions of the proposed USMCA establish new rules and disciplines to remove trade barriers and counter discriminatory action while also providing governments with flexibility. The provisions go much further than the KORUS agreement in establishing obligations on multiple aspects of digital trade, and contrast sharply with China's authoritarian approach discussed above.

USMCA provisions prohibit customs duties and discrimination against digital products, requirements for source code or algorithms disclosure, or technology transfer mandates. The agreement protects electronic authentication and signatures, electronic payment systems, and consumer access to the Internet. Provisions require anti-spam measures, domestic legal frameworks for online consumer and personal privacy protection, and identifies specific key principles and international guidelines that the parties must take into account. USMCA contains broad provisions to protect cross-border data flows and restrict data localization requirements;

---

[165] KORUS FTA, Chapter 13, Annex 13-B, Section B, https://ustr.gov/sites/default/files/uploads/agreements/fta/korus/asset_upload_file35_12712.pdf.

[166] The USMCA text is available at https://ustr.gov/trade-agreements/free-trade-agreements/united-states-mexicocanada-agreement.

[167] For more information on the TPP, see CRS In Focus IF10000, *TPP: Overview and Current Status*, by Brock R. Williams and Ian F. Fergusson.

for financial services, open data flows is subject to the financial regulator having access to data necessary to fulfill its regulatory and supervisory role. The digital trade chapter also prohibits liability of internet intermediaries, in line with current U.S. law, and promotes the publication of government data through open-data formats. The parties agree to cooperate on and promote a number of issues including risk-based cybersecurity, privacy, SMEs, and the APEC Cross-Border Privacy Rules (see below).

OTHER INTERNATIONAL FORUMS FOR DIGITAL TRADE

Given the cross-cutting nature of the digital world, digital trade issues touch on other policy objectives and priorities, such as privacy and national security. While U.S. and international trade agreements are one way for the United States to establish market opening and new rules and disciplines to govern digital trade, not every issue is necessarily suitable for an international trade agreement and not every international partner is ready, or willing, to take on such commitments. In other international forums outside of trade negotiations, other tools can be used to encourage high-level, nonbinding best practices and principles and align expectations.

G-20

The influential Group of 20 (G-20) is one venue for establishing common principles, and digital issues have been on its agenda recently.[168] At the 2017 meeting, G-20 leaders established the Digital Economy Task Force (DETF). The G-20 Digital Economy Ministerial Meeting issued a

[168] The Group of Twenty (G-20) is a forum for advancing international cooperation and coordination among 20 major advanced and emerging-market economies. The G-20 includes Argentina, Australia, Brazil, Canada, China, France, Germany, India, Indonesia, Italy, Japan, Mexico, Russia, Saudi Arabia, South Africa, South Korea, Turkey, United Kingdom, and the United States, as well as the European Union (EU). For more information on the G-20, see CRS Report R40977, *The G-20 and International Economic Cooperation: Background and Implications for Congress*, by Rebecca M. Nelson.

declaration that identified requisites for a thriving digital economy and specific recommendations.[169] As host, Japan is expected to build on the digital economy agenda in 2019, with a specific emphasis on privacy and data governance.

OECD

The OECD provides a forum to discuss principles and norms to facilitate a thriving digital economy. The OECD issued a series of reports in 2017 and 2018 related to digital trade, including an assessment of the digital transformation of each OECD economy[170] and bridging the digital gender divide.[171] The reports identified specific challenges and recommendations, including establishing a national digital strategy and removing market access barriers. The United States could work with its OECD partners to reinforce principles, including an open Internet and the need to balance public policy objectives. The OECD Global Forum on the Digital Security for Prosperity also allows for multi-stakeholder international engagement to discuss issues such as the governance of digital security issues.

APEC

The Asian Pacific Economic Cooperation (APEC) forum presents another opportunity for sharing best practices and setting high-level principles on issues that may be of greater concern to developing countries

[169] https://g20.argentina.gob.ar/en/news/g20-confirms-importancc-digital-economy-global-development.

[170] OECD, *Key Issues for Digital Transformation in the G20*, January 12, 2017, https://www.oecd.org/internet/keyissues-for-digital-transformation-in-the-g20.pdf.

[171] OECD, *Bridging the Digital Gender Divide: Include, upskill, innovate*, October 30, 2018, http://www.oecd.org/sti/ ieconomy/bridging-the-digital-gender-divide.pdf.

with less advanced digital economies and industry.[172] APEC is implementing the Cross-Border Privacy Rules (CBPR) system to be consistent with the already established APEC Privacy Framework.[173] According to the Business Software Alliance, most countries across the globe have data protection frameworks based on either the APEC CBPR system or the EU regime, but some countries still lack privacy laws.[174] Currently, the United States, Japan, Mexico, Canada, South Korea, Singapore, Taiwan, and Australia are CBPR members; the Philippines is in the process of joining. Some observers view CBPR, which aims to reflect a diversity of national privacy regimes, as a scalable solution that could potentially be adopted multilaterally. Others may view the EU regime as a more comprehensive, top-down approach. Due to its voluntary nature, APEC has served as an incubator for potential plurilateral agreements.

Regulatory Cooperation

Ongoing regulatory cooperation efforts are another important tool for addressing differences between parties, better aligning regulatory requirements, and reducing inconsistencies and redundancies that can hamper or discriminate against the free flow of data, goods, and services. These forums provide an opportunity for U.S. agencies to work directly with overseas counterparts and focus on specific aspects of digital trade such as online privacy, consumer protection, and rules for online contract formation and enforcement. The EU-U.S. Privacy Shield is one example of regulatory authorities working together to address such issues.

---

[172] Asia Pacific Economic Cooperation (APEC) is a regional economic forum established in 1989 with 21 Asian Pacific economies as members. See http://www.apec.org/About-Us/About-APEC.aspx.

[173] http://publications.apec.org/Publications/2011/10/Enabling-Electronic-Commerce-The-Contribution-of-APECsData-Privacy-Framework.

[174] http://cloudscorecard.bsa.org/2018/index.html; http://cloudscorecard.bsa.org/2016/.

ISSUES FOR CONGRESS

Policy questions continue to evolve as the internet-driven economy and innovations grow. Digital trade is intimately connected to and woven into all parts of the U.S. economy and overlaps with other sectors, requiring policymakers to balance many different objectives. For example, digital trade relies on cross-border data flows, but policymakers must balance open data flows with public policy goals such as protecting privacy, supporting law enforcement, and improving personal and national security and safety.

The complexity of the debate related to cross-border data flows and digital trade more generally involves complementary and competing interests and stakeholders. Companies and individuals who seek to do business abroad, and trade negotiators who seek to open markets may focus on maintaining open market access, which may include cross-border data flows, while others may want to limit foreign competition. Privacy advocates may focus on protecting personal information. Meanwhile, law enforcement and defense advisors may seek the ability to access or limit information flows based on national security interests.

Digital trade raises numerous complex issues of potential interest to Congress with possible legislative and oversight implications. Issues include

- Understanding of the economic impact of digital trade on the U.S. economy and the effects of localization and other digital trade barriers on U.S. exports, jobs, and competition.
- Examining how best to balance market openness and cross-border data flows with other policy goals, such as right to privacy and the government's need for access to protect safety and national security.
- Considering if the United States would benefit from overarching digital privacy policy and what lessons can be drawn from other countries' experiences, and how to best balance this with U.S. trade negotiating objectives.

- Effectively addressing important digital trade barriers and cybertheft.
- Considering how best to assure public confidence and trust in network reliability and security that underlie the global digital economy and allow it to effectively and efficiently function.
- Examining evolving U.S. trade policy efforts, including how the proposed USMCA, WTO plurilateral, and potential new bilateral negotiations may address U.S. trade barriers, set new rules and disciplines, and respond to different standard-setting practices that may have global reach, including by the EU and China.
- Assessing if U.S. agencies have the necessary tools to accurately measure the size and scope of digital trade in order to analyze the impact of potential policies.
- Assessing the effectiveness of the Trump Administration's Section 301 actions involving Chinese trade practices and other bilateral efforts related to cybersecurity and digital trade.

APPENDIX. DIGITAL TRADE BARRIERS

Barriers to Internet Services

- Discriminatory treatment of digital goods and services
- Duties on digital goods or services
- Foreign investment restrictions
- Intermediary liability without safe harbor or fair-use provisions that could make internet platforms responsible for content posted by users
- Low de minimis threshold for customs duties on imported goods, including e-commerce purchases
- "Snippet tax" on search engines that quote text snippets as part of search results

- Taxes on over-the-top (OTT) services such as media, messaging, or voice-overinternet-protocol (VOIP)
- Web filtering and blocking of content

Localization Barriers

- Data localization requirements prohibiting cross-border data flows and requiring the use of local servers for data storage or processing
- Limited or no access to foreign government procurement markets
- Requirement for use of local technology
- Comprehensive privacy regulations that may discriminate against foreign providers

Technology Barriers

- Restrictions or prohibitions on use of encryption
- Source code, technology, or other intellectual property rights (IPR) forced transfer requirements
- Local testing and certification for imported information technology (IT) equipment may add costs or delays for imported goods

Other Barriers

- Cybersecurity threats or local requirements
- Weak IPR enforcement

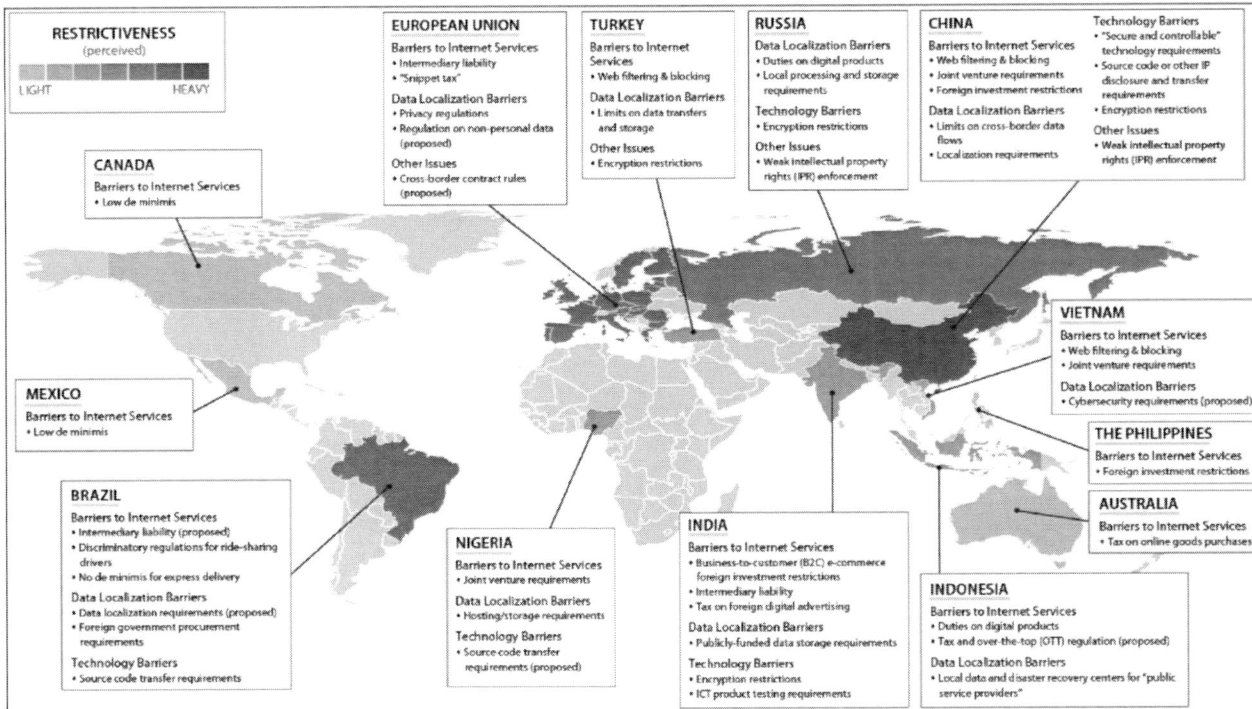

Source: CRS based on U.S. Trade Representative, *2018 National Trade Estimate Report on Foreign Trade Barriers*. Note: This map is illustrative of digital trade barriers and not meant to be an exhaustive list.

Figure A.1. Levels of Perceived Digital Trade Barriers in Selected Countries (according to the U.S. Trade Representative).

In: Key Congressional Reports for May 2019 ISBN: 978-1-53616-335-3
Editor: Aston Dunkley

Chapter 3

FEDERAL RESEARCH AND DEVELOPMENT (R&D) FUNDING: FY2020*

John F. Sargent Jr., Tadlock Cowan, Robert Esworthy, Marcy E. Gallo, Laurie A. Harris, Eva Lipiec, Daniel Morgan and Kavya Sekar

ABSTRACT

President Trump's budget request for FY2020 includes approximately $134.1 billion for research and development (R&D). Several FY2019 appropriations bills had not been enacted at the time the President's FY2020 budget was prepared; therefore, the President's budget included the FY2018 actual funding levels, 2019 annualized continuing resolution (CR) levels, and the FY2020 request levels. On February 15, 2019, Congress enacted the Consolidated Appropriations Act, 2019 (P.L. 116-6). This act included each of the remaining appropriations acts, completing the FY2019 appropriations process. The act also rendered the CR levels identified in the budget no longer relevant, though for some agencies the exact amount of R&D funding in

* This is an edited, reformatted and augmented version of Congressional Research Service, Publication No. R45715 dated May 9, 2019.

the act remained uncertain. The analysis of government-wide R&D funding in this chapter compares the President's request for FY2020 to the FY2018 level. For agencies for which the FY2019 R&D funding levels are known, individual agency analyses in this chapter compare the FY2020 request to FY2019 enacted levels. For agencies for which the FY2019 R&D funding levels remain unknown, individual agency analyses in this chapter compare the FY2020 request to FY2018 actual levels. As of the date of this chapter, the House had not completed action on any of the 12 regular appropriations bills for FY2020; nor had the Senate.

In FY2018, OMB adopted a change to the definition of development, applying a more narrow treatment it describes as "experimental development." This change was intended to harmonize the reporting of U.S. R&D funding data with the approach used by other nations. The new definition is used in this chapter. Under the new definition of R&D (applied to both FY2018 and FY2020 figures), President Trump is requesting approximately $134.1 billion for R&D for FY2020, a decrease of $1.7 billion (1.2%) from the FY2018 level. Adjusted for inflation, the President's FY2020 R&D request represents a decrease of 5.1% below the FY2018 level.

Funding for R&D is concentrated in a few departments and agencies. In FY2018, eight federal agencies received 96.3% of total federal R&D funding, with the Department of Defense (DOD, 38.6%) and the Department of Health and Human Services (HHS, 27.2%) combined accounting for nearly two-thirds of all federal R&D funding. The same eight agencies account for 97.2% of the FY2020 request, with DOD accounting for 44.3% and HHS for 25.1%.

Under the President's FY2020 budget request, most federal agencies would see their R&D funding decline. The primary exception is the Department of Defense. DOD's requested R&D funding for FY2020 is $7.1 billion (13.5%) above the FY2018 level. The Departments of Transportation and Veterans Affairs would see small increases in R&D funding. Among the agencies with the largest proposed reductions in R&D funding in the FY2020 budget compared to the FY2018 actual levels are the Department of Energy ($2.8 billion, 15.8%), the National Science Foundation ($567 million, 9.0%), and National Aeronautics and Space Administration ($475 million, 4.0%).

The President's FY2020 budget request would reduce funding for basic research by $1.5 billion (4.0%), applied research by $4.3 billion (10.5%), and facilities and equipment by $0.5 billion (12.8%), while increasing funding for development by $4.5 billion (8.3%).

President Trump's FY2020 budget is largely silent on funding levels for multiagency R&D initiatives. However, some activities supporting these initiatives are discussed in agency budget justifications and are reported in the agency analyses in this chapter.

The request represents the President's R&D priorities. Congress may opt to agree with none, part, or all of the request, and it may express different priorities through the appropriations process. In recent years, Congress has completed the annual appropriations process after the start of the fiscal year. Completing the process after the start of the fiscal year and the accompanying use of continuing resolutions can affect agencies' execution of their R&D budgets, including the delay or cancellation of planned R&D activities and the acquisition of R&D-related equipment.

INTRODUCTION

The 116th Congress continues its interest in U.S. research and development (R&D) and in evaluating support for federal R&D activities. The federal government has played an important role in supporting R&D efforts that have led to scientific breakthroughs and new technologies, from jet aircraft and the internet to communications satellites, shale gas extraction, and defenses against disease. In recent years, widespread concerns about the federal debt, recent and projected federal budget deficits, and federal budget caps have driven difficult decisions about the prioritization of R&D, both in the context of the entire federal budget and among competing needs within the federal R&D portfolio. Increases in the budget caps for FY2018 and FY2019 reduced some of the pressure affecting these decisions, but the concerns remain and the caps for FY2020 have not been increased.

The U.S. government supports a broad range of scientific and engineering R&D. Its purposes include specific concerns such as addressing national defense, health, safety, the environment, and energy security; advancing knowledge generally; developing the scientific and engineering workforce; and strengthening U.S. innovation and competitiveness in the global economy. Most of the R&D funded by the federal government is performed in support of the unique missions of individual funding agencies.

The federal R&D budget is an aggregation of the R&D activities of these agencies. There is no single, centralized source of R&D funds. Agency R&D budgets are developed internally as part of each agency's overall budget development process. R&D funding may be included either in accounts that are entirely devoted to R&D or in accounts that include funding for non-R&D activities. Agency budgets are subjected to review, revision, and approval by the Office of Management and Budget (OMB) and become part of the President's annual budget submission to Congress. The federal R&D budget is then calculated by aggregating the R&D activities of each federal agency.

Congress plays a central role in defining the nation's R&D priorities as it makes decisions about the level and allocation of R&D funding—overall, within agencies, and for specific programs. In recent years, some Members of Congress have expressed concerns about the level of federal spending (for R&D and for other purposes) in light of the federal deficit and debt. Other Members of Congress have expressed support for increased federal spending for R&D as an investment in the nation's future competitiveness. As Congress acts to complete the FY2020 appropriations process, it faces two overarching issues: the amount of the federal budget to be spent on federal R&D and the prioritization and allocation of the available funding.

This chapter begins with a discussion of the overall level of R&D in President Trump's FY2020 budget request, followed by analyses of R&D funding in the request from a variety of perspectives and for selected multiagency R&D initiatives. The remainder of the report discusses and analyzes the R&D budget requests of selected federal departments and agencies that, collectively, account for approximately 99% of total federal R&D funding.

Selected terms associated with federal R&D funding are defined in the text box on the next page. Appendix A provides a list of acronyms and abbreviations.

Definitions Associated with Federal Research and Development Funding

Two key sources of definitions associated with federal research and development funding are the White House Office of Management and Budget and the National Science Foundation.

- **Office of Management and Budget.** The Office of Management and Budget provides the following definitions of R&D-related terms in OMB Circular No. A-11, "Preparation, Submission, and Execution of the Budget."[1] This document provides guidance to agencies in the preparation of the President's annual budget and instructions on budget execution. In 2017, OMB adopted a refinement to the categories of R&D, replacing "development" with "experimental development," which more narrowly defines the set of activities to be included. The new definition has resulted in lower reported R&D by some agencies, including the Department of Defense and the National Aeronautics and Space Administration. This definition is used in the President's FY2020 budget.
- **Conduct of R&D**. Research and experimental development (R&D) activities are defined as creative and systematic work undertaken in order to increase the stock of knowledge—including knowledge of people, culture, and society—and to devise new applications using available knowledge.
- **Basic Research**. Basic research is defined as experimental or theoretical work undertaken primarily to acquire new knowledge of the underlying foundations of phenomena and observable facts. Basic research may include activities with broad or general applications in mind, such as the study of how plant genomes change, but should exclude research directed towards a specific application or requirement, such as the optimization of the genome of a specific crop species.
- **Applied Research**. Applied research is defined as original investigation undertaken in order to acquire new knowledge. Applied research is, however, directed primarily towards a specific practical aim or objective.
- **Experimental Development**. Experimental development is defined as creative and systematic work, drawing on knowledge gained from research and practical experience, which is directed at producing new products or processes or improving existing products or processes. Like research, experimental development will result in gaining additional knowledge.
- **R&D Equipment.** R&D equipment includes amounts for major equipment for research and development. It includes acquisition, design, or production of major movable equipment, such as mass spectrometers, research vessels, DNA sequencers, and other major movable instruments for use in R&D activities. It includes programs of $1 million or more that are devoted to the purchase or construction of major R&D equipment.

[1] The White House, Office of Management and Budget, Circular No. A-11, "Preparation, Submission, and Execution of the Budget," June 2018, https://www.whitehouse.gov/wp-content/uploads/2018/06/a11.pdf.

(Continued)

- **R&D Facilities**. R&D facilities includes amounts for the construction of facilities that are necessary for the execution of an R&D program. This may include land, major fixed equipment, and supporting infrastructure such as a sewer line or housing at a remote location.
- **National Science Board/National Science Foundation**. The National Science Board/National Science Foundation provides the following definitions of R&D-related terms in its *Science and Engineering Indicators: 2018* report.[2]
- **Research and Development (R&D):** Research and experimental development comprise creative and systematic work undertaken to increase the stock of knowledge—including knowledge of humankind, culture, and society—and its use to devise new applications of available knowledge.
- **Basic Research:** Experimental or theoretical work undertaken primarily to acquire new knowledge of the underlying foundations of phenomena and observable facts, without any particular application or use in view.
- **Applied Research:** Original investigation undertaken to acquire new knowledge; directed primarily, however, toward a specific, practical aim or objective.
- **Experimental Development:** Systematic work, drawing on knowledge gained from research and practical experience and producing additional knowledge, which is directed to producing new products or processes or to improving existing products or processes.

THE PRESIDENT'S FY2020 BUDGET REQUEST

On March 11, 2019, President Trump released his proposed FY2020 budget. He provided additional details the following week.

Completion of the FY2019 budget process on February 15, 2019, more than four months after the start of FY2019, as well as a government shutdown, led to both a delay in the scheduled release of the President's FY2020 budget request, and the use by the Office of Management and Budget of a mix of estimated continuing appropriations act FY2019 funding levels (generally, for agencies whose FY2019 appropriations were enacted after the start of FY2019)[3] and enacted FY2019 funding levels (generally, for agencies whose appropriations were enacted prior to the

[2] National Science Board/National Science Foundation, *Science and Engineering Indicators 2018*, January 2018, https://www.nsf.gov/statistics/2018/nsb20181/.

[3] Agencies that received their FY2019 appropriations in the Consolidated Appropriations Act, 2019 (P.L. 116-6).

start of FY2019).[4] As a result, the aggregate (total) FY2019 R&D funding levels for all agencies in the *Analytical Perspectives* addendum to the President's FY2020 budget are estimated "using FY 2019 enacted appropriations where available and annualized Continuing Resolution [levels] for agencies without enacted appropriations prior to Feb. 15, 2019."[5] With enactment of the remaining FY2019 appropriations acts in the Consolidated Appropriations Act, 2019 (P.L. 116-6), the Administration's estimated aggregate R&D funding level no longer accurately reflects total enacted FY2019 R&D funding. OMB has not issued a document with comprehensive R&D figures for each agency or in aggregate.

Therefore, the analysis of government-wide R&D funding in this chapter, as well as of some of the individual agency analyses, compares the President's request for FY2020 to the FY2018 actual level.

President Trump is proposing approximately $134.1 billion for R&D for FY2020, a decrease of $1.7 billion (1.2%) below the FY2018 level of $135.8 billion. Adjusted for inflation, the President's FY2020 R&D request represents a constant-dollar decrease of 5.2% from the FY2018 actual level.[6]

The President's request includes continued R&D funding for existing single-agency and multiagency programs and activities, as well as new initiatives. This chapter provides government-wide, multiagency, and individual agency analyses of the President's FY2020 request as it relates to R&D and related activities. Additional information and analysis will be

---

[4] Agencies that received their FY2019 appropriations in the Energy and Water, Legislative Branch, and Military Construction and Veterans Affairs Appropriations Act, 2019 (P.L. 115-244), and the Department of Defense and Labor, Health and Human Services, and Education Appropriations Act, 2019 and Continuing Appropriations Act, 2019 (P.L. 115-245).

[5] Executive Office of the President (EOP), OMB, *Analytical Perspectives, Budget of the United States Government, Fiscal Year 2020, Research and Development,* March 18, 2019, p. 269, https://www.whitehouse.gov/wp-content/ uploads/2019/03/ap_21_research-fy2020.pdf.

[6] As calculated by CRS using the Gross Domestic Product (GDP) (chained) price index for FY2018 and FY2020 in Table 10.1, "Gross Domestic Product and Deflators Used in the Historical Tables: 1940–2022," *Budget of the United States Government, Fiscal Year 2019,* https://www.whitehouse.gov/wp-content/uploads/2018/02/hist10z1-fy2019.xlsx.

included as the House and Senate act on the President's budget request through appropriations bills.

Other Factors Affecting Analysis of the FY2019 Budget Request

Other factors complicate the analysis of changes in R&D funding for FY2019, both in aggregate and for selected agencies. For example:

- Beginning in FY2018, OMB replaced the R&D category "development" with a subset referred to as "experimental development" in an effort that OMB asserts better aligns its data with the survey data collected by the National Science Foundation, and for greater consistency with international standards. The new definition excludes some development activities, primarily at the Department of Defense (DOD) and the National Aeronautics and Space Administration (NASA), that had been characterized as development in previous budgets. The new definition (experimental development) is used throughout this chapter for FY2017 and later years, except in the section "Department of Defense," which treats non-experimental development funding as part of its total research, development, testing, and evaluation budget in line with DOD practice.
- Inconsistency among agencies in the reporting of R&D and the inclusion of R&D activities in accounts with non-R&D activities may result in different figures being reported by OMB and the White House Office of Science and Technology Policy (OSTP), including those shown in Table 1, and those in agency budget analyses that appear later in this chapter.

FEDERAL R&D FUNDING PERSPECTIVES

Federal R&D funding can be analyzed from a variety of perspectives that provide different insights. The following sections examine the data by agency, by the character of the work supported, and by a combination of these two perspectives.

Federal R&D by Agency

Congress makes decisions about R&D funding through the authorization and appropriations processes primarily from the perspective of individual agencies and programs.

Table 1 provides data on R&D funding by agency for FY2018 (actual), FY2019 (enacted, for selected agencies), and FY2020 (request).[7] Enacted data for FY2019 is provided only for agencies whose FY2019 appropriations process was completed before the FY2020 budget request was finalized.

Under the request, eight federal agencies would receive more than 97% of total federal R&D funding in FY2020: the Department of Defense, 44.3%; Department of Health and Human Services (HHS), primarily the National Institutes of Health (NIH), 25.1%; Department of Energy (DOE), 11.0%; National Aeronautics and Space Administration, 8.4%; National Science Foundation (NSF), 4.3%; Department of Agriculture (USDA), 1.8%; Department of Commerce (DOC), 1.3%; and Department of Veterans Affairs (VA), 1.0%. This chapter provides an analysis of the R&D budget requests for these agencies, as well as for the Department of Homeland Security (DHS), Department of the Interior (DOI), Department of Transportation (DOT), and Environmental Protection Agency (EPA).

All but three federal agencies would see their R&D funding decrease under the President's FY2020 request compared to FY2018. The only agencies with increased R&D funding in FY2020 relative to the FY2018 level would be DOD (up $7.077 billion, 13.5%), VA (up $39 million, 3.0%), and DOT (up $33 million, 3.2%). The agencies with largest R&D funding declines in the FY2020 request compared to FY2018 (as measured in dollars) would occur in the budgets of HHS (down $3.249 billion, 8.8%), DOE (down $2.764 billion, 15.8%), NSF (down $567 million, 29.7%), and NASA (down $475 million, 4.0%).

Among agencies for which FY2019 enacted funding is shown in the President's budget, FY2020 R&D funding would increase only for DOD (up $3.631 billion, 6.5%). HHS R&D funding would decline by $4.954 billion (12.8%). DOE R&D funding would decline by $3.075 billion (17.3%). VA R&D funding would decline by $17 million (1.3%).

---

[7] EOP, OMB, *Analytical Perspectives, Budget of the United States Government, Fiscal Year 2020, Research and Development,* March 18, 2019, p. 269, https://www.whitehouse.gov/wp-content/uploads/2019/03/ap_21_researchfy2020.pdf.

Department of Education R&D funding would decline by $34 million (13.2%). See Table 1.

Table 1. Federal Research and Development Funding by Agency, FY2018-FY2020 (budget authority, dollar amounts in millions)

Department/ Agency	FY2018 Actual	FY2019 Enacted	FY2020 Request	Change, FY2018-FY2020		Change, FY2019-FY2020	
				Dollar	Percent, Total	Dollar	Percent, Total
Department of Defense	$52,386	$55,832	$59,463	$7,077	13.5%	$3,631	6.5%
Dept. of Health and Human Services	36,942	38,647	33,693	-3,249	-8.8%	-4,954	-12.8%
Department of Energy	17,482	17,793	14,718	-2,764	-15.8%	-3,075	-17.3%
NASA	11,755	n/a	11,280	-475	-4.0%	n/a	n/a
National Science Foundation	6,327	n/a	5,760	-567	-9.0%	n/a	n/a
Department of Agriculture	2,618	n/a	2,464	-154	-5.9%	n/a	n/a
Department of Commerce	2,029	n/a	1,694	-335	-16.5%	n/a	n/a
Department of Veterans Affairs	1,286	1,342	1,325	39	3.0%	17	-1.3%
Department of Transportation	1,043	n/a	1,076	33	3.2%	n/a	n/a
Department of the Interior	885	n/a	753	-132	-14.9%	n/a	n/a
Department of Homeland Security	725	n/a	507	-218	-30.1%	n/a	n/a
Environmental Protection Agency	492	n/a	285	-207	-42.1%	n/a	n/a
Smithsonian Institution	357	n/a	315	-42	-11.8%	n/a	n/a
Department of Education	257	258	224	-33	-12.8%	34	-13.2%
Other	1,181	n/a	540	-641	-54.3%	n/a	n/a
Total	135,765	n/a	134,097	-1,668	-1.2%	n/a	n/a

Source: CRS analysis of data from EOP, OMB, *Analytical Perspectives, Budget of the United States Government, Fiscal Year 2020, Research and Development,* March 18, 2019, p. 269, https://www.whitehouse.gov/wp-content/uploads/2019/03/ap_21_research-fy2020.pdf. The "FY2019" column includes enacted funding levels for only those agencies that received appropriations prior to February 15, 2019, for which OMB provided estimated levels of R&D based on the enacted legislation.

Notes: Components may not sum to totals due to rounding. n/a=not available.

Federal R&D by Character of Work, Facilities, and Equipment

Federal R&D funding can also be examined by the character of work it supports—basic research, applied research, or development—and by funding provided for construction of R&D facilities and acquisition of major R&D equipment. (See Table 2). President Trump's FY2020 request includes $35.164 billion for basic research, down $1.452 billion (4.0%) from FY2018; $40.707 billion for applied research, down $4.264 billion (10.5%); $59.108 billion for development, up $4.543 billion (8.3%); and $3.382 billion for facilities and equipment, down $495 million (12.8%).

Table 2. Federal R&D Funding by Character of Work and Facilities and Equipment, FY2018 and FY2020 (budget authority, dollar amounts in millions)

Character of Work, Facilities, and Equipment	FY2018 Actual	FY2020 Request	Change, FY2018-FY2020	
			Dollars	Percent
Basic research	36,616	35,164	-1,452	-4.0%
Applied research	40,707	36,443	-4,264	-10.5%
Development	54,565	59,108	4,543	8.3%
Facilities and Equipment	3,877	3,382	-495	-12.8%
Total	135,765	134,097	-1,668	-1.2%

Source: CRS analysis of data from EOP, OMB, *Analytical Perspectives, Budget of the United States Government, Fiscal Year 2020, Research and Development,* March 18, 2019, pp. 271-272, https://www.whitehouse.gov/wpcontent/uploads/2019/03/ap_21_research-fy2020.pdf.

Note: Components may not sum to totals due to rounding.

Federal Role in U.S. R&D by Character of Work

A primary policy justification for public investments in basic research and for incentives (e.g., tax credits) for the private sector to conduct research is the view, widely held by economists, that the private sector will, left on its own, underinvest in basic research from a societal perspective. The usual argument for this view is that the social returns (i.e., the benefits to society at large) exceed the private returns (i.e., the benefits accruing to the private investor, such as increased revenues or higher stock value). Other factors that may inhibit corporate investment in basic research include long time horizons for achieving commercial applications (diminishing the potential returns due to the time value of money), high levels of technical risk/uncertainty, shareholder demands for shorter-term returns, and asymmetric and imperfect information.

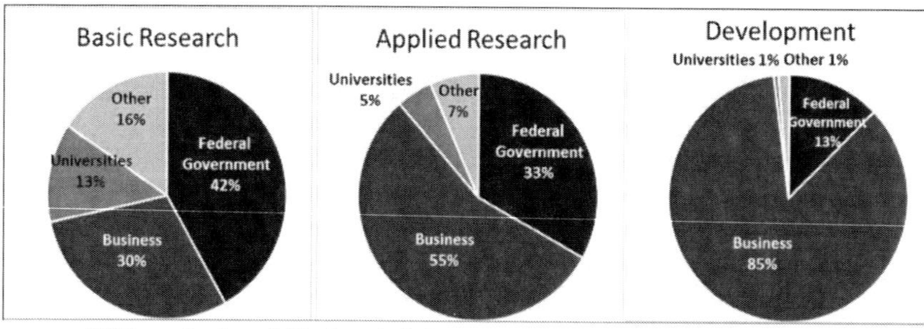

Source: CRS analysis of National Science Foundation, National Patterns of R&D Resources: 2016–17 Data Update, NSF 19-309, February 27, 2019, https://ncses.nsf.gov/pubs/nsf19309/.

Notes: Components may not add to total due to rounding. Data for 2017 is estimated and subject to revision.

Figure 1. Composition of U.S. Basic Research, Applied Research, and Development by Funding Sector, 2017.

The federal government is the nation's largest supporter of basic research, funding 42% of U.S. basic research in 2017. Business funded 30% of U.S. basic research in 2017, with state governments, universities, and other nonprofit organizations funding the remaining 29%. For U.S. applied research, business is the primary funder, accounting for an estimated 55% in 2017, while the federal government accounted for an estimated 33%. State governments, universities, and other nonprofit organizations funded the remaining 12%. Business also provides the vast majority of U.S. funding for development. Business accounted for 85% of development funding in 2017, while the federal government provided 13%.

State governments, universities, and other nonprofit organizations funded the remaining 2% (see Figure 1).[8]

Table 3. Selected R&D Funding Agencies by Character of Work, Facilities, and Equipment, FY2018 Actual and FY2020 Request (budget authority, dollar amounts in millions)

Character of Work/Agency	FY2018 Actual	FY2020 Request	Change, FY2018-FY2020	
			Dollars	Percent
Basic Research				
Health and Human Services	18,278	16,785	-1,493	-8.2%
Energy	5,005	4,647	-358	-7.2%
National Science Foundation	5,066	4,568	-498	-9.8%
Applied Research				
Health and Human Services	18,414	16,624	-1,790	-9.7%
Energy	7,998	6,410	-1,588	-19.9%
Defense	5,690	5,440	-250	-4.4%
Experimental Development				
Defense	44,363	51,686	7,323	16.5%
NASA	5,872	3,791	-2,081	-35.4%
Energy	2,549	1,952	-597	-23.4%
Facilities and Equipment				
Energy	1,930	1,709	-221	-11.5%
Commerce	633	566	-67	-10.6%
National Science Foundation	503	505	2	0.4%

Source: CRS analysis of data from EOP, OMB, Analytical Perspectives, Budget of the United States Government, Fiscal Year 2020, Research and Development, March 18, 2019, pp. 271-272, https://www.whitehouse.gov/wp-content/ uploads/2019/03/ap_21_research-fy2020.pdf.

Note: This table shows only the top three funding agencies in each category, based on the FY2020 request.

Federal R&D by Agency and Character of Work Combined

Federal R&D funding can also be viewed from the combined perspective of each agency's contribution to basic research, applied research, development, and facilities and equipment. Table 3 lists the three

[8] CRS analysis of National Science Foundation, *National Patterns of R&D Resources: 2016–17 Data Update*, NSF 19- 309, February 27, 2019, https://ncses.nsf.gov/pubs/nsf19309/. Components may not add to total due to rounding.

agencies with the most funding in each of these categories as proposed in the President's FY2020 budget. The overall federal R&D budget reflects a wide range of national priorities, including supporting advances in spaceflight, developing new and affordable sources of energy, and understanding and deterring terrorist groups. These priorities and the mission of each individual agency contribute to the composition of that agency's R&D spending (i.e., the allocation among basic research, applied research, development, and facilities and equipment). In the President's FY2020 budget request, the Department of Health and Human Services, primarily NIH, would account for nearly half (47.7%) of all federal funding for basic research. HHS would also be the largest federal funder of applied research, accounting for about 45.6% of all federally funded applied research in the President's FY2020 budget request. DOD would be the primary federal funder of development, accounting for 87.4% of total federal development funding in the President's FY2020 budget request. DOE would be the primary federal funder of facilities and equipment, accounting for 50.5% of total federal facilities and equipment funding in the President's FY2020 budget request.[9]

MULTIAGENCY R&D INITIATIVES

For many years, presidential budgets have reported on multiagency R&D initiatives. Often, they have also provided details of agency funding for these initiatives. Some of these efforts have a statutory basis—for example, the Networking and Information Technology Research and Development (NITRD) program, the National Nanotechnology Initiative (NNI), and the U.S. Global Change Research Program (USGCRP). These programs generally produce annual budget supplements identifying objectives, activities, funding levels, and other information, usually published shortly after the presidential budget release. Other multiagency

[9] EOP, OMB, *Analytical Perspectives, Budget of the United States Government, Fiscal Year 2019,* February 12, 2018, pp. 238-240, https://www.whitehouse.gov/omb/budget/Analytical_Perspectives.

R&D initiatives have operated at the discretion of the President without such a basis and may be eliminated at the discretion of the President. President Trump's FY2020 budget is largely silent on funding levels for these efforts and whether any or all of the nonstatutory initiatives will continue. Some activities related to these initiatives are discussed in agency budget justifications and may be addressed in the agency analyses later in this chapter. This section provides available multiagency information on these initiatives.

Networking and Information Technology Research and Development Program (NITRD)

Established by the High-Performance Computing Act of 1991 (P.L. 102-194), the Networking and Information Technology Research and Development program is the primary mechanism by which the federal government coordinates its unclassified networking and information technology R&D investments in areas such as supercomputing, high-speed networking, cybersecurity, software engineering, and information management. In FY2019, 21 agencies are NITRD members; nonmember agencies also participate in NITRD activities. NITRD efforts are coordinated by the National Science and Technology Council (NSTC) Subcommittee on Networking and Information Technology Research and Development.[10]

P.L. 102-194, as reauthorized by the American Innovation and Competitiveness Act of 2017 (P.L. 114-329), requires the director of

[10] The NSTC was established by Executive Order 12881 in 1993. According to the White House, "This Cabinet-level Council is the principal means within the Executive Branch to coordinate science and technology policy across the diverse entities that make up the Federal research and development enterprise. Chaired by the President, the membership of the NSTC is made up of the Vice President, Cabinet Secretaries and Agency Heads with significant science and technology responsibilities, and other White House officials. In practice, the Assistant to the President for Science and Technology Policy oversees the NSTC's ongoing activities." (Source: The White House, Office of Science and Technology Policy, "NSTC," https://www.whitehouse.gov/ostp/nstc/). For more information on the NSTC, see CRS Report R43935, *Office of Science and Technology Policy (OSTP): History and Overview*, by John F. Sargent Jr. and Dana A. Shea; and https://www.whitehouse.gov/ostp/nstc/.

NITRD to prepare an annual report to be delivered to Congress along with the President's budget request. This annual report is to include, among other things, detailed information on the program's budget for the current and previous fiscal years, and the proposed budget for the next fiscal year. The latest annual report was published in August 2018 and related to the FY2019 budget request. President Trump requested $5,277.6 million for NITRD research in FY2019. In FY2017, NITRD funding was $5,126.4 million. The President's FY2020 budget does not identify aggregate funding levels for NITRD for FY2018 (actual), FY2019 (estimate), or FY2020 (request). The FY2020 NSF budget request includes $1.20 billion in NITRD funding for FY2020, a decrease of $98.2 million (7.6%) from FY2018 (actual).

Table 4. Networking and Information Technology Research and Development Program Funding, FY2017-FY2020 (budget authority, in millions of current dollars)

FY2017 Actual	FY2018 Actual	FY2019 Request	FY2020 Request
$5,126.4	TBD	$5,277.6	TBD

Source: Subcommittee on Networking and Information Technology Research and Development, Committee on Technology, National Science and Technology Council, The White House, *The Networking and Information Technology Research and Development Program: Supplement to the President's FY2019 Budget,* August 2018.

Notes: The President's FY2020 budget does not identify aggregate funding levels for NITRD for FY2018 (actual), FY2019 (enacted), or FY2020 (request). TBD = To be determined.

U.S. Global Change Research Program (USGCRP)

The U.S. Global Change Research Program coordinates and integrates federal research and applications to understand, assess, predict, and respond to human-induced and natural processes of global change. The program seeks to advance global climate change science and to "build a knowledge base that informs human responses to climate and global change through coordinated and integrated Federal programs of research, education, communication, and decision support."[11] In FY2018, 13 departments and agencies participated in the USGCRP. USGCRP efforts are coordinated by the NSTC Subcommittee on Global Change Research. Each agency develops and carries out its activities as its contribution to the USGCRP, and funds are appropriated to each agency for those activities; those activities may or may not be identified as associated with the USGCRP in the President's annual budget proposal or each agency's budget justification to Congress.

The Global Change Research Act of 1990 (GCRA) (P.L. 101-606) requires each federal agency or department involved in global change research to report annually to Congress on each element of its proposed global change research activities, as well as the portion of its budget request allocated to each element of the program.[12] The President is also required to identify those activities and the annual global change research budget in the President's annual budget request to Congress. Some, but not all, participating agencies provide the required information in their budget justifications.

In addition, in almost each of the past 17 years, language in appropriations laws has required the President to submit a comprehensive report to the appropriations committees "describing in detail all Federal agency funding, domestic and international, for climate change programs, projects, and activities ... including an accounting of funding by agency...."[13] The President's most recent report was submitted in January 2017 for the FY2017 request.

The President's FY2020 budget does not identify aggregate funding levels for USGCRP for FY2018 (actual), FY2019 (estimate), or the FY2020 (request). Without budget reports pursuant to the GCRP Act or appropriations laws, the budget authority or expenditures publicly reported by agencies is generally insufficient to independently build a complete and

---

[11] U.S. Global Change Research Program website, http://www.globalchange.gov/about/mission-vision-strategic-plan..

[12] Directives to report annually to Congress on budget requests and spending occur in several sections of P.L. 101-606, including Sections 105(b) and (c) on Budget Coordination, and Section 107, Annual Report.

[13] See, most recently, P.L. 115-31, Consolidated Appropriations Act, 2017, Section 416.

consistent estimate of funding for the USGCRP. Below are results of CRS research for selected agencies, which incompletely represent federal funding in recent years that support the USGCRP.

Table 5. U.S. Global Change Research Program Funding in Selected Agencies, FY2017-FY2020 (budget authority, in millions of current dollars)

	FY2016 Enacted	FY2017 Enacted	FY2018 Enacted	FY2019 Enacted	FY2020 Request
National Aeronautics and Space Admin.	1,534	1,555	1,555	TBD	TBD
Department of Energy (Office of Science)	238	229	239	259	117
National Oceanic and Atmospheric Admin.	308	338	TBD	TBD	TBD
Office of Atmospheric Research					
Climate Research Program	157	156	157	159	88
National Science Foundation	339	243	254	n/a	224
Environmental Protection Agency	19	22	17	19	0
Total, USGCRP[a]	2,599	2,790	TBD	TBD	TBD

Source: Office of Management and Budget, "Federal Expenditures on Science, Energy, and International Assistance Programs that Advance the Federal Response to Climate Change," submitted to the House Committee on Energy and Commerce, January 19, 2017. Note that these figures are more recent than and differ slightly from the totals reported in the most recent report from the USGRCP, *Our Changing Planet FY2017*, p. 42. https://www.globalchange.gov/browse/reports/our-changing-planet-fy-2017. For FY2017 enacted to FY2020 request: Congressional Budget Justification documents of each department or agency for each fiscal year request. NASA communicated its estimates to CRS. Estimates for EPA come from the section entitled "Eliminated Programs."

Notes: The President's FY2020 budget does not identify aggregate funding levels for USGCRP for FY2018 (actual), FY2019 (enacted), or FY2020 (request). TBD = To be determined. Funding for the USGCRP is appropriated to more than a dozen federal departments and agencies, and some spending of it is transferred or coordinated through interagency agreements. Almost all of the funding is spent directly by agencies on research and related activities; a small percentage is spent for interagency coordination and communications in the USGCRP program office. Italicized numbers are components of an agency's total and are not added separately in the total.

[a] The reported total ("Total, USGCRP") includes funding for agencies not included in the table.

The President's FY2020 budget does not identify aggregate funding levels for USGCRP for FY2018 (actual), FY2019 (estimate), or the FY2020 (request). Without budget reports pursuant to the GCRP Act or appropriations laws, the budget authority or expenditures publicly reported by agencies is generally insufficient to independently build a complete and consistent estimate of funding for the USGCRP. Table 5 presents results of CRS research for selected agencies, which incompletely represent federal funding in recent years that supports the USGCRP.

National Nanotechnology Initiative (NNI)

Launched in FY2001, the National Nanotechnology Initiative is a multiagency R&D initiative to advance understanding and control of matter at the nanoscale, where the physical, chemical, and biological properties of materials differ in fundamental and useful ways from the properties of individual atoms or bulk matter.[14] In 2003, Congress enacted the 21st Century Nanotechnology Research and Development Act (P.L. 108-153), providing a legislative foundation for some of the activities of the NNI. NNI efforts are coordinated by the NSTC Subcommittee on Nanoscale Science, Engineering, and Technology (NSET). In FY2019, the President's request included NNI funding for 16 federal departments and independent agencies and commissions with budgets dedicated to nanotechnology R&D. The NSET includes other federal departments and independent agencies and commissions with responsibilities for health, safety, and environmental regulation; trade; education; intellectual property; international relations; and other areas that might affect or be affected by nanotechnology.

P.L. 108-153 requires the NSTC to prepare an annual report to be delivered to Congress at the time the President's budget request is sent to Congress. This annual report is to include detailed information on the program's budget for the current fiscal year and the program's proposed budget for the next fiscal year, as well as additional information and data related to the performance of the program. The latest annual report was published in August 2018. President Trump requested $1,395.6 million for

---

[14] In the context of the NNI and nanotechnology, the nanoscale refers to lengths of 1 to 100 nanometers. A nanometer is one-billionth of a meter, or about the width of 10 hydrogen atoms arranged side by side in a line.

NNI research in FY2019. In FY2017, NNI funding was $1,552.3 million.[15] The President's FY2020 budget does not identify aggregate funding levels for NNI for FY2018 (actual), FY2019 (enacted), or FY2020 (request). The NSF budget request includes $389 million in NNI funding for FY2020, $179 million (31.4%) less than in FY2018.

Table 6. National Nanotechnology Initiative Funding, FY2017-FY2019 (budget authority, in millions of current dollars)

FY2017 Actual	FY2018 Actual	FY2019 Request	FY2020 Request
$1,552.3	TBD	$1,395.6	TBD

Source: Subcommittee on Nanoscale Science, Engineering, and Technology, Committee on Technology, National Science and Technology Council, The White House, *The National Nanotechnology Initiative: Supplement to the President's 2019 Budget*, August 2018.

Notes: The FY2018 estimate in the FY2019 supplement ($1,477.4 million) was based on annualized 2018 continuing resolution levels; the actual FY2018 number will be based on final FY2018 appropriations and agency expenditures. The President's FY2020 budget does not identify aggregate funding levels for NNI for FY2018 (actual), FY2019 (enacted), or FY2020 (request).

American Artificial Intelligence Initiative

In February 2019, President Trump signed Executive Order 13859 establishing an American Artificial Intelligence Initiative to accelerate national leadership in artificial intelligence (AI).[16] Among other things, the EO directs the heads of implementing agencies that perform or fund R&D to consider AI as an agency R&D priority, in accordance with their missions and consistent with applicable law. In particular, the EO directs the Secretaries of Defense, Commerce, Health and Human Services, and Energy, the Administrator of the National Aeronautics and Space Administration, and the Director of the National Science Foundation to prioritize the allocation of high-performance computing resources for AI-related applications through increased assignment of discretionary funding

---

[15] EOP, National Science and Technology Council, *The National Nanotechnology Initiative: Supplement to the President's 2019 Budget*, August 2018.

[16] Executive Order 13859, "Maintaining American Leadership in Artificial Intelligence," 84 *Federal Register* 3967- 3972, February 11, 2019, https://www.federalregister.gov/documents/2019/02/14/2019-02544/maintaining-americanleadership-in-artificial-intelligence.

and any other appropriate mechanisms. According to *Analytical Perspectives,* the President's FY2020 budget would provide approximately $850 million for this initiative to the Department of Energy, National Institutes of Health, National Institute of Standards and Technology, and National Science Foundation.[17]

National Quantum Initiative

In December 2018, President Trump signed the National Quantum Initiative Act (P.L. 115-368) establishing a National Quantum Initiative with the stated purpose of ensuring "the continued leadership of the United States in quantum information science [QIS] and its technology applications." The act requires the establishment of a 10-year plan to accelerate the development of QIS and technology applications. According to *Analytical Perspectives*, the President's FY2020 budget includes approximately $430 million for this initiative at DOD, DOE, NIST, and NSF.[18]

Other Initiatives

A number of presidential initiatives without statutory foundations were in operation at the end of the Obama Administration, but have not been addressed explicitly in President Trump's FY2018, FY2019, or FY2020 budgets. Two of these are part of the Advanced Manufacturing Partnership (AMP): the National Robotics Initiative (NRI) and Manufacturing USA (formerly known as the National Network for Manufacturing Innovation or NNMI).[19]

---

[17] EOP, OMB, *Analytical Perspectives, Budget of the United States Government, Fiscal Year 2020, Research and Development,* March 18, 2019, pp. 269, https://www.whitehouse.gov/wp-content/uploads/2019/03/ap_21_researchfy2020.pdf.

[18] Ibid., p. 270.

[19] Funding for Manufacturing USA is requested in the National Institute of Standards and Technology (NIST) and Department of Defense budget justifications. NIST's FY2020 budget justification requests $15.2 million to support the National Institute for Innovation in Manufacturing Biopharmaceuticals (NIIMBL, approximately $10 million), which was competed and selected by NIST, and for the Manufacturing USA network (approximately $5 million). In addition, the Department of Defense is requesting $66.2 million for FY2020

According to *Analytical Perspectives*, the President's FY2020 budget prioritizes R&D aimed at advances in manufacturing and the integration of those advances into the domestic supply chain to reduce U.S. reliance on foreign sources of critical products. Budget priorities include intelligent manufacturing systems, materials and processing technologies, advances in semiconductor design and fabrication, and innovations in food and agricultural manufacturing.[20]

Other initiatives include the Cancer Moonshot, the Brain Research through Advancing Innovative Neurotechnologies (BRAIN) Initiative, the Precision Medicine Initiative (PMI), the Materials Genome Initiative (MGI), and an effort to double federal funding for clean energy R&D. Some of the activities of these initiatives are discussed in agency budget justifications and the agency analyses later in this chapter.

FY2020 Appropriations Status

The remainder of this chapter provides a more in-depth analysis of R&D in 12 federal departments and agencies that, in aggregate, receive nearly 99% of total federal R&D funding. Agencies are presented in order of the size of their FY2020 R&D budget requests, with the largest presented first.

Annual appropriations for these agencies are provided through 9 of the 12 regular appropriations bills. For each agency covered in this chapter, Table 7 shows the corresponding regular appropriations bill that provides primary funding for the agency, including its R&D activities.

---

for the DOD-sponsored institutes that are part of the Manufacturing USA network as part of the request of the Office of the Secretary of Defense (see DOD program element 0603680D8Z).

[20] EOP, OMB, *Analytical Perspectives, Budget of the United States Government, Fiscal Year 2020, Research and Development,* March 18, 2019, p. 270.

Table 7. Alignment of Agency R&D Funding and Regular Appropriations Bills

Department/Agency	Regular Appropriations Bill
Department of Defense	Department of Defense Appropriations Act
Department of Health and Human Services -National Institutes of Health	(1) Departments of Labor, Health and Human Services, and Education, and Related Agencies Appropriations Act (2) Department of the Interior, Environment, and Related Agencies Appropriations Act
Department of Energy	Energy and Water Development and Related Agencies Appropriations Act
National Aeronautics and Space Administration	Commerce, Justice, Science, and Related Agencies Appropriations Act
National Science Foundation	Commerce, Justice, Science, and Related Agencies Appropriations Act
Department of Agriculture	Agriculture, Rural Development, Food and Drug Administration, and Related Agencies Appropriations Act
Department of Commerce -National Institute of Standards and Technology -National Oceanic and Atmospheric Administration	Commerce, Justice, Science, and Related Agencies Appropriations Act
Department of Veterans Affairs	Military Construction and Veterans Affairs, and Related Agencies Appropriations Act
Department of the Interior	Department of the Interior, Environment, and Related Agencies Appropriations Act
Department of Transportation	Transportation, Housing and Urban Development, and Related Agencies Appropriations Act
Department of Homeland Security	Department of Homeland Security Appropriations Act
Environmental Protection Agency	Department of the Interior, Environment, and Related Agencies Appropriations Act

Source: CRS Report R40858, Locate an Agency or Program Within Appropriations Bills, by Justin Murray.

Because of the way that agencies report budget data to Congress, it can be difficult to identify the portion that is R&D. Consequently, R&D data presented in the agency analyses in this chapter may differ from R&D data in the President's budget or otherwise provided by OMB.

Funding for R&D is often included in appropriations line items that also include non-R&D activities; therefore, in such cases, it may not be possible to identify precisely how much of the funding provided in appropriations laws is allocated to R&D specifically. In general, R&D

funding levels are known only after departments and agencies allocate their appropriations to specific activities and report those figures.

As of the date of this chapter, the House had completed action on none of the 12 regular appropriations bills for FY2020; the Senate had completed action on none of the bills. None of the 12 had been enacted as law.

DEPARTMENT OF DEFENSE[21]

The mission of the Department of Defense is to provide "the military forces needed to deter war and ensure our nation's security."[22] Congress supports research and development activities at DOD primarily through the department's Research, Development, Test, and Evaluation (RDT&E) funding. These funds support the development of the nation's future military hardware and software and the science and technology base upon which those products rely.

Most of what DOD spends on RDT&E is appropriated in Title IV of the annual defense appropriations bill. (See Table 8). However, RDT&E funds are also appropriated in other parts of the bill. For example, RDT&E funds are appropriated as part of the Defense Health Program, the Chemical Agents and Munitions Destruction Program, and the National Defense Sealift Fund. The Defense Health Program (DHP) supports the delivery of health care to DOD personnel and their families. DHP funds (including the RDT&E funds) are requested through the Defense-wide Operations and Maintenance appropriations request.

---

[21] This section was written by John F. Sargent Jr., Specialist in Science and Technology Policy, CRS Resources, Science, and Industry Division.

[22] Department of Defense, https://www.defense.gov/Our-Story/.

Table 8. Department of Defense RDT&E (total obligational authority[a], in millions of dollars)

Budget Account	FY2019 Enacted	FY2020 Request			FY2020 House			FY2020 Senate			FY2020 Enacted		
	Base +OCO	Base	OCO	Total	Base	OCO	Total	Base	OCO	Total	Base	OCO	Total
Army	11,375.2	12,192.8	204.1	12,396.9									
Navy	18,657.8	20,270.5	164.1	20,434.9									
Air Force	41,488.6	45,616.1	450.2	46,066.4									
Defense-wide	24,062.0	24,347.0	828.0	25,174.9									
Director, Operational Test and Evaluation	377.0	221.2	0.0	221.2									
Total Title IV—By Account Budget Activity													
6.1 Basic Research	2,528.4	2,320.0	0.0	2,320.0									
6.2 Applied Research	6,025.1	5,315.5	1.7	5,317.2									
6.3 Advanced Technology Development	7,105.6	6,423.0	74.8	6,497.8									
6.4 Advanced Component Development and Prototypes	22,005.0	26,722.2	215.0	26,937.2									
6.5 Systems Dev. and Demonstration	15,577.2	17,652.3	113.0	17765.3									
6.6 Management Supportb	7,152.0	6,769.5	1.9	6,771.4									
6.7 Operational Systems Developmentc	35,567.2	37,445.0	1,240.3	38,685.3									
Total Title IV—by Budget Activity	95,960.6	102,647.5	1,646.7	104,294.3									
Title V—Revolving and Management Funds													
National Defense Sealift Fund	0.0	0.0	0.0	0.0									

Table 8. (Continued)

Budget Account	FY2019 Enacted	FY2020 Request			FY2020 House			FY2020 Senate			FY2020 Enacted		
	Base +OCO	Base	OCO	Total	Base	OCO	Total	Base	OCO	Total	Base	OCO	Total
Title VI—Other Defense Programs													
Defense Health Program	2,179.6	732.3	0.0	732.3									
Chemical Agents and Munitions Destruction	886.7	875.9	0.0	875.9									
Inspector General	4.0	3.0	0.0	3.0									
Grand Total	99,030.9	104,258.7	1,646.7	105,905.4									

Source: CRS analysis of Department of Defense Budget, Fiscal Year 2020, RDT&E Programs (R-1), March 2019. Notes: Figures for the columns headed "FY2019 House," "FY2019 Senate" and "FY2019 Enacted" will be added, if available, as each action is completed. Totals may differ from the sum of the components due to rounding.

[a]According to DOD, "Total Obligation Authority (TOA) is the sum of: (1) all budget authority (BA) granted (or requested) from the Congress in a given year, (2) amounts authorized to be credited to a specific fund, 3) BA transferred from another appropriation, and 4) Unobligated balances of BA from previous years which remain available for obligation. In practice, this term is used primarily in discussing the DoD budget, and most often refers to TOA as the 'direct program,' which equates to only (1) and (2) above." DOD defines "budget authority" as "the authority becoming available during the year to enter into obligations that result in immediate or future outlays of Government funds." See DoD 7000.14-R, "Department of Defense Financial Management Regulation," http://comptroller.defense.gov/fmr.aspx.

[b]Includes funding for Director of Test and Evaluation.

[c]Includes funding for Classified Programs.

The program's RDT&E funds support congressionally directed research on breast, prostate, and ovarian cancer; traumatic brain injuries; orthotics and prosthetics; and other medical conditions. Congress appropriates funds for this program in Title VI (Other Department of Defense Programs) of the defense appropriations bill. The Chemical Agents and Munitions Destruction Program supports activities to destroy the U.S. inventory of lethal chemical agents and munitions to avoid future risks and costs associated with storage. Funds for this program are requested through the Defense-wide Procurement appropriations request. Congress appropriates funds for this program also in Title VI. The National Defense Sealift Fund supports the procurement, operation and maintenance, and research and development associated with the nation's naval reserve fleet and supports a U.S. flagged merchant fleet that can serve in time of need. In some fiscal years, RDT&E funding for this effort is requested in the Navy's Procurement request and appropriated in Title V (Revolving and Management Funds) of the appropriations bill. RDT&E funds also have been requested and appropriated as part of DOD's separate funding to support efforts in what the George W. Bush Administration termed the Global War on Terror (GWOT) and what the Obama and Trump Administrations have referred to as Overseas Contingency Operations (OCO). In appropriations bills, the term Overseas Contingency Operations/Global War on Terror (OCO/GWOT) has been used; President Trump's FY2020 budget uses the term Overseas Contingency Operations. Typically, the RDT&E funds appropriated for OCO activities go to specified Program Elements (PEs) in Title IV.

According to the Comptroller of the Department of Defense, the FY2020 OCO request is divided into three requirement categories—direct war, enduring, and OCO for base.[23] For purposes of this chapter, these categories of OCO funding requests will be reported collectively.

In addition, OCO/GWOT-related requests/appropriations have included money for a number of transfer funds. In the past, these have included the Iraqi Freedom Fund (IFF), the Iraqi Security Forces Fund, the Afghanistan Security Forces Fund, and the Pakistan Counterinsurgency

[23] "Direct War Requirements" reflect combat or combat support costs that are not expected to continue once combat operations end at major contingency locations. "OCO for Enduring Requirements" reflects enduring in-theater and CONUS costs that will remain after combat operations end. "OCO for Base Requirements" reflects funding for base budget requirements, which support the National Defense Strategy, such as defense readiness, readiness enablers, and munitions, financed in the OCO budget to comply with the base budget defense caps included in current law. Department of Defense, Office of the Under Secretary of Defense (Comptroller)/Chief Financial Officer, *Defense Budget Overview: Department of Defense Fiscal Year 2020 Budget Request*, March 19, 2019, https://comptroller.defense.gov/Portals/45/Documents/defbudget/fy2020/fy2020_Budget_Request_Overview_Book.pdf.

Capability Fund. Congress typically has made a single appropriation into each such fund and authorized the Secretary to make transfers to other accounts, including RDT&E, at his discretion. These transfers are eventually reflected in Title IV prior-year funding figures.

For FY2020, the Trump Administration is requesting $104.294 billion for DOD's Title IV RDT&E PEs (base plus OCO), $8.334 billion (8.7%) above the enacted FY2019 level. (See Table 8). In addition, the request includes $732.3 million in RDT&E through the Defense Health Program (DHP; down $1.447 billion, 66.4% from FY2019), $875.9 million in RDT&E through the Chemical Agents and Munitions Destruction program (down $10.8 million, 1.2% from FY2019), and $3.0 million for the Inspector General for RDT&E-related activities (down $1.0 million, 25.4% from FY2019). The FY2020 budget included no RDT&E funding via the National Defense Sealift Fund, the same as the FY2019 enacted level.

RDT&E funding can be analyzed in different ways. RDT&E funding can be characterized organizationally. Each of military department requests and receives its own RDT&E funding. So, too, do various DOD agencies (e.g., the Missile Defense Agency and the Defense Advanced Research Projects Agency), collectively aggregated within the Defense-Wide account. RDT&E funding also can be characterized by budget activity (i.e., the type of RDT&E supported). Those budget activities designated as 6.1, 6.2, and 6.3 (basic research, applied research, and advanced technology development, respectively) constitute what is called DOD's Science and Technology program (S&T) and represent the more research-oriented part of the RDT&E program. Budget activities 6.4 and 6.5 focus on the development of specific weapon systems or components for which an operational need has been determined and an acquisition program established. Budget activity 6.6 provides management support, including support for test and evaluation facilities. Budget activity 6.7 supports the development of system improvements in existing operational systems.[24] Many congressional policymakers are particularly interested in DOD S&T

[24] For additional information on the structure of Defense RDT&E, see CRS Report R44711, *Department of Defense Research, Development, Test, and Evaluation (RDT&E): Appropriations Structure*, by John F. Sargent Jr.

program funding since these funds support the development of new technologies and the underlying science. Some in the defense community see ensuring adequate support for S&T activities as imperative to maintaining U.S. military superiority into the future. The knowledge generated at this stage of development may also contribute to advances in commercial technologies. The FY2020 request for Title IV S&T funding (base plus OCO) is $14.135 billion, $1.524 billion (9.7%) below the FY2019 enacted level. Within the S&T program, basic research (6.1) receives special attention, particularly by the nation's universities. DOD is not a large supporter of basic research when compared to NIH or NSF. However, over half of DOD's basic research budget is spent at universities. The Trump Administration is requesting $2.320 billion for DOD basic research for FY2020, $208.4 million (8.2%) below the FY2019 enacted level. DOD is a substantial source of federal funds for university R&D in certain fields, such as aerospace, aeronautical, and astronautical engineering (40%); electrical, electronic, and communications engineering (39%); mechanical engineering (28%); computer and information sciences (28%); and materials science (25%).[25]

DEPARTMENT OF HEALTH AND HUMAN SERVICES

The mission of the Department of Health and Human Services (HHS) is "to enhance and protect the health and well-being of all Americans ... by providing for effective health and human services and fostering advances in medicine, public health, and social services."[26] This section focuses on HHS research and development funded through the National Institutes of Health, an HHS agency that accounts for nearly 97% of total HHS R&D funding.[27] Other HHS agencies that provide funding for R&D include the

---

[25] CRS analysis of data from NSF, *Higher Education Research and Development Survey, Fiscal Year 2017,* data tables, November 20, 2018, https://ncsesdata.nsf.gov/herd/2017/.

[26] U.S. Department of Health and Human Services, "About," http://www.hhs.gov/about.

[27] CRS analysis of data from OMB, *Analytical Perspectives, Budget of the United States Government, Fiscal Year 2020,* 2019, p. 268, https://www.whitehouse.gov/wp-content/uploads/2019/03/spec-fy2020.pdf.

Centers for Disease Control and Prevention (CDC), Centers for Medicare and Medicaid Services (CMS), Food and Drug Administration (FDA), Agency for Healthcare Research and Quality (AHRQ), Health Resources and Services Administration (HRSA), and Administration for Children and Families (ACF); additional R&D funding is attributed to departmental management.[28]

National Institutes of Health[29]

NIH is the primary agency of the federal government charged with performing and supporting biomedical and behavioral research. It also has major roles in training biomedical researchers and disseminating health information. The NIH mission is "to seek fundamental knowledge about the nature and behavior of living systems and the application of that knowledge to enhance health, lengthen life, and reduce illness and disability."[30] The agency consists of the NIH Office of the Director (OD) and 27 institutes and centers (ICs).

The OD sets overall policy for NIH and coordinates the programs and activities of all NIH components, particularly in areas of research that involve multiple institutes. The ICs focus on particular diseases (e.g., National Cancer Institute), areas of human health and development (e.g., National Institute on Aging), or scientific research fields or support (e.g., National Center for Advancing Translational Sciences). Each IC plans and manages its own research programs in coordination with OD. As shown in Table 9, separate appropriations are provided to 24 of the 27 ICs, as well as to OD, the Innovation Account (established by the 21st Century Cures Act, P.L. 114-255), and an intramural Buildings and Facilities account. The other three centers, which perform centralized support services, are funded through the other ICs.

---

[28] Ibid.

[29] This section was written by Kavya Sekar, Analyst in Health Policy, CRS Domestic Social Policy Division. For background information on NIH, see CRS Report R41705, *The National Institutes of Health (NIH): Background and Congressional Issues*, by Judith A. Johnson and Kavya Sekar.

[30] HHS, National Institutes of Health, "About NIH, What We Do, Mission and Goals," http://www.nih.gov/about-nih/what-we-do/mission-goals.

NIH supports and conducts a wide range of basic and clinical research, research training, and health information dissemination across all fields of biomedical and behavioral sciences. According to NIH, about 10% of the NIH budget supports intramural research projects conducted by the nearly 6,000 NIH federal scientists, most of whom are located on the NIH campus in Bethesda, MD. More than 80% of NIH's budget goes to the extramural research community in the form of grants, contracts, and other awards. This funding supports research performed by more than 300,000 nonfederal scientists and technical personnel who work at more than 2,500 universities, hospitals, medical schools, and other research institutions.[31]

Funding for NIH comes primarily from the annual Labor, HHS, and Education (LHHS) appropriations act, with an additional amount for Superfund-related activities from the Interior/Environment appropriations act. Those two appropriations acts provide NIH's discretionary budget authority. In addition, NIH has received mandatory funding of $150 million annually provided in the Public Health Service Act (PHSA), Section 330B, for a special program on type 1 diabetes research.[32] Some funding is also transferred to NIH pursuant to the "PHS Evaluation Tap" Transfer authority.[33] The total funding available for NIH activities, taking account of add-ons and PHS tap transfers, is known as the NIH program level.[34] President Trump's FY2020 budget request would provide NIH a total program level of $34.368 billion, a decrease of $4.941 billion (12.6%) compared with FY2019 enacted levels.[35] The proposed FY2020 program level total would include $33.410 billion provided through LHHS appropriations (including the full amount authorized by the 21st Century Cures Act); $741 million from the PHS evaluation transfer; $66.581

---

[31] NIH, "What We Do: Budget," https://www.nih.gov/about-nih/what-we-do/budget.

[32] 42 U.S.C. §254c-2.

[33] For more information on the PHS evaluation tap, or PHS Evaluation Set-Aside, see discussion in CRS Report R44916, *Public Health Service Agencies: Overview and Funding (FY2016-FY2018)*, coordinated by C. Stephen Redhead and Agata Dabrowska.

[34] Totals include amounts "transferred in" pursuant to PHS tap, but do not include any amounts "transferred out" under this same authority.

[35] Based on CRS analysis using budget data from NIH, "FY2020 Justification of Estimates for Appropriations Committees—Overview, Vol. 1," p. 78, and the National Institute of Environmental Health Sciences appropriation from the Consolidated Appropriations Act, 2019 (P.L. 116-6).

million provided through Interior/Environment appropriations for Superfund-related activities; and $150 million in proposed funding for the mandatory type 1 diabetes program. Under the FY2020 budget proposal, all existing ICs and budget activity lines, except for Buildings and Facilities, would receive a decrease compared to FY2019 enacted levels (see Table 9).[36] The Buildings and Facilities appropriation of $200 million would not change from FY2019 to FY2020.

Additionally, the FY2020 Budget Request proposes consolidating the Agency for Healthcare Research and Quality into NIH, forming a 28th IC—the National Institute for Research on Safety and Quality (NIRSQ). The creation of a new NIH institute would require an amendment to PHSA Section 401(d), which specifies that "[i]n the National Institutes of Health, the number of national research institutes and national centers may not exceed a total of 27." Under the request, NISRQ would receive $256 million in funding for FY2020.[37] The main funding mechanism NIH uses to support extramural research is research project grants (RPGs), which are competitive, peer-reviewed, and largely investigator-initiated. Historically, over 50% of the NIH budget is used to support RPGs, which include salaries for investigators and research staff.[38] The FY2020 budget proposes to reduce the average cost of RPGs by capping the percentage of an investigator's salary that can be paid with grant funds to 90%.[39] The FY2020 Trump budget proposal includes $150 million in mandatory

---

[36] Though the budget request provides an increase to the National Institute of General Medical Sciences through discretionary LHHS budget authority, the total amount for NIGMS with the PHS evaluation transfer included is less than FY2019 enacted levels. For proposed FY2020 IC funding levels see NIH, "FY2020 Justification of Estimates for Appropriations Committees—Overview, Vol. 1," p. 78.

[37] The FY2020 request for NISRQ is roughly $82 million less than the discretionary amount appropriated to Agency for Healthcare Research and Quality (AHRQ) in FY2019. In addition, AHRQ also received roughly $112.5 million in mandatory funds in FY2019 from the Patient-Centered Outcomes Research Trust Fund (PCORTF). Funding for the PCORTF is scheduled to expire at the end of FY2019. The FY2020 President's budget does not request any new mandatory funds for the PCORTF. For more information, see "FY2020 Justification of Estimates for Appropriations Committees—NIH National Institute for Research on Safety and Quality," p. 6.

[38] CRS analysis of data from NIH, *Actual Total Obligations by Budget Mechanism: FY2000-FY2018*, March 18, 2019, https://officeofbudget.od.nih.gov/pdfs/FY20/Mechanism-Detail-for-NIH-FY-2000-FY-2018.pdf.

[39] NIH, "FY2020 Justification of Estimates for Appropriations Committees—Overview, Vol. 1," p. 13.

funding for research on type 1 diabetes authorized under the PHS Act Section 330B within the budget of the National Institute of Diabetes and Digestive and Kidney Diseases (NIDDK). For this proposal, Congress and the President would need to enact legislation to extend the special diabetes program funding, because under current law, no new funding will be available for this program after September 30, 2019.[40] Additionally, the FY2020 program level request includes $741 million in funding transferred to NIH by the PHS evaluation tap. Discretionary funding for certain programs at NIH and other HHS agencies that are authorized under the PHS Act can be subject to an assessment under Section 241 of the PHS Act (42 U.S.C. §238j). This provision authorizes the Secretary to use a portion of eligible appropriations to study the effectiveness of federal health programs and to identify improvements. Although the PHS Act limits the tap to no more than 1% of eligible appropriations, in recent years, annual LHHS appropriations acts have specified a higher amount (2.5% in FY2019)[41] and have also typically directed specific amounts of funding from the tap for transfer to a number of HHS programs. The assessment has the effect of redistributing appropriated funds for specific purposes among PHS and other HHS agencies. NIH, with the largest budget among the PHS agencies, has historically been the largest "donor" of program evaluation funds; until recently, it had been a relatively minor recipient.[42] Provisions in recent LHHS appropriations acts have directed specific tap transfers to NIH, making NIH a net recipient of tap funds.

The FY2020 NIH budget request also includes $492 million made available through the 21st Century Cures Act (see text box below; hereinafter referred to as "The Cures Act"). The Cures Act (P.L. 114-255) specifies that $149 million is for the Precision Medicine Initiative, $140 million is for the BRAIN Initiative, $195 million is for cancer research, and $8 million is for research on regenerative medicine for FY2020.[43]

---

[40] 42 U.S.C. §254c-2.

[41] P.L. 115-245, §204.

[42] For more information, see the "PHS Evaluation Set-Aside" section of CRS Report R44916, Public Health Service Agencies: Overview and Funding (FY2016-FY2018), coordinated by C. Stephen Redhead and Agata Dabrowska. By convention, budget tables such as Table 9 do not subtract the amount of the evaluation tap from the donor agencies' appropriations.

[43] P.L. 114-255, Sec. 1001(b)(4).

The President's FY2020 budget identifies several research priorities for NIH in the coming year. The overview below outlines some of these priority themes in the budget request.

Confronting the Opioids Crisis

The request includes $1.3 billion designated for opioids and pain research across NIH, with $500 million of the total for the Helping to End Addiction Long-Term (HEAL) initiative. The HEAL Initiative, launched in April 2018, aims to accelerate the development of new medications and devices to treat opioid addiction. In addition, NIH plans to support research on neonatal abstinence syndrome, chronic pain, and other opioids-related issues.[44]

The 21st Century Cures Act and the NIH Innovation Account

The 21st Century Cures Act (P.L. 114-255) created the NIH Innovation account and specified that funds in the account must be appropriated in order to be available for expenditure. The Cures Act specified that the following amounts shall be transferred to the NIH Innovation account: $352 million for FY2017; $496 million for FY2018;
$711 million for FY2019; $492 million for FY2020; $404 million for FY2021; $496 million for FY2022; $1,085 million for FY2023; $407 million for FY2024; $127 million for FY2025; and $226 million for FY2026.[45]

The Cures Act authorizes four projects (with total funding from FY2017 to FY2026): the Precision Medicine Initiative ($1.455 billion), the BRAIN Initiative ($1.511 billion), cancer research ($1.8 billion), and regenerative medicine using adult stem cells ($30 million). Amounts, once appropriated, are to be available until expended. The NIH Director may transfer these amounts from the NIH Innovation account to other NIH accounts but only for the purposes specified in the Cures Act. If the NIH Director determines that the funds for any of the four Innovation Projects are not necessary, the amounts may be transferred back to the NIH Innovation account. This transfer authority is in addition to other transfer authorities provided by law.

---

[44] HHS, "Budget in Brief—FY2020," pp. 53-54, https://www.hhs.gov/sites/default/files/fy-2020-budget-in-brief.pdf.

[45] The first round of funding was provided by Section 194 of the Further Continuing and Security Assistance Appropriations Act, 2017 (CR, P.L. 114-254). The CR appropriated $352 million in the NIH Innovation account for necessary expenses to carry out the four NIH Innovation Projects as described in Section 1001(b)(4) of the Cures Act. The second round of funding was provided by the FY2018 omnibus appropriations act (P.L. 115-141). The third round of funding ($711 million) is provided by the FY2019 Consolidated Defense, LHHS, and Continuing Resolution appropriations act (P.L. 115-245).

Pediatric Research

The FY2020 request proposes $50 million in designated funding for a pediatric cancer initiative. The initiative, designed to complement existing pediatric cancer research, would aggregate data on pediatric cancer cases and coordinate existing datasets to create a "comprehensive, shared resource to support research on childhood cancer in all its forms."[46] The request also designates $15 million for the Institutional Development Award (IDeA) States Pediatric Clinical Trials Network to support pediatric clinical studies, such as on the "dosing, safety, and efficacy of drugs that are commonly prescribed to children."[47]

Supporting the Next Generation of Researchers

The request would provide $100 million in dedicated funding for the Next Generation Researchers Initiative, which aims to support new and early stage scientists in attaining their first NIH grants.[48] Through the program, NIH ICs are to create funding pathways and other strategies targeted at new and early-stage scientists, and would be required to collect data and evaluate their outcomes.[49]

Ending the HIV Epidemic

As a part of a proposed HHS wide plan, "Ending the HIV Epidemic: A Plan for America," the FY2020 request designates $6 million in funding to Centers for AIDS Research to collect data and inform HHS on best practices for the initiative. The goal for the plan is to reduce new infections by 75% in the next 5 years, and by 90% in the next 10 years.[50]

---

[46] HHS, "Budget in Brief—FY2020," p. 55, https://www.hhs.gov/sites/default/files/fy-2020-budget-in-brief.pdf.

[47] Ibid, p. 56. The Institutional Development Award (IDeA) program is designed to build research capacity in states that have had historically low levels of NIH funding. Currently, 23 states and Puerto Rico are eligible for IDeA funding. See National Institute of General Medical Sciences (NIGMS), "Institutional Development Award," https://www.nigms.nih.gov/Research/DRCB/IDeA/Pages/default.aspx, and CRS Report R44689, *Established Program to Stimulate Competitive Research (EPSCoR): Background and Selected Issues*, by Laurie A. Harris.

[48] HHS, "Budget in Brief—FY2020," p. 56, https://www.hhs.gov/sites/default/files/fy-2020-budget-in-brief.pdf.

[49] NIH, "Next Generation Researchers Initiative," https://grants.nih.gov/ngri.htm.

[50] Ibid, p. 56.

New Technologies and Biomedical Research

Table 9. National Institutes of Health Funding (budget authority, in millions of dollars)

Institutes/Centers	FY2019 Enacted	FY2020 Request	FY2020 House	FY2020 Senate	FY2020 Enacted
Cancer Institute (NCI)	$6,144	$5,247			
Heart, Lung, and Blood Institute (NHLBI)	$3,488	$3,003			
Dental/Craniofacial Research (NIDCR)	$462	$397			
Diabetes/Digestive/Kidney (NIDDK)[a]	$2,030	$1,747			
Neurological Disorders/Stroke (NINDS)	$2,274	$2,026			
Allergy/Infectious Diseases (NIAID)	$5,523	$4,754			
General Medical Sciences (NIGMS)[b]	$1,726	$1,732			
Child Health/Human Development (NICHD)	$1,506	$1,297			
National Eye Institute (NEI)	$797	$686			
Environmental Health Sciences (NIEHS)[c]	$775	$667			
National Institute on Aging (NIA)	$3,083	$2,654			
Arthritis/Musculoskeletal/Skin Diseases (NIAMS)	$605	$521			
Deafness/Communication Disorders (NIDCD)	$474	$408			
National Institute of Mental Health (NIMH)	$1,870	$1,630			
National Institute on Drug Abuse (NIDA)	$1,420	$1,296			
Alcohol Abuse/Alcoholism (NIAAA)	$526	$452			
Nursing Research (NINR)	$163	$140			
Human Genome Research Institute (NHGRI)	$576	$495			
Biomedical Imaging/Bioengineering (NIBIB)	$389	$335			
Minority Health/Health Disparities (NIMHD)	$315	$271			
Complementary/Integrative Health (NCCIH)	$146	$126			
Advancing Translational Sciences (NCATS)	$806	$694			
Fogarty International Center (FIC)	$78	$67			
National Library of Medicine (NLM)	$442	$380			
Office of Director (OD)[d]	$1,917	$1,769			
Innovation Accounte	$196	$157			
Buildings and Facilities (B&F)	$200	$200			
National Institute for Research on Safety & Quality (NIRSQ)[f]	–	$256			
Subtotal, NIH	$37,933	$33,410			
PHS Program Evaluation	$1,147	$741			
Superfund (Interior approp. to NIEHS)[g]	$79	$67			

Institutes/Centers	FY2019 Enacted	FY2020 Request	FY2020 House	FY2020 Senate	FY2020 Enacted
Mandatory type 1 diabetes funds[h]	$150	$150			
NIH Program Level	$39,308	$34,368			

Source: NIH, "FY2020 Justification of Estimates for Appropriations Committees-Vol 1, Overview," p. 78, except as noted below.

Notes: Totals may differ from the sum of the components due to rounding. Amounts in table may differ from actuals in many cases. By convention, budget tables such as Table 9 do not subtract the amount of transfers to the evaluation tap from the agencies' appropriation. Figures for the columns headed "FY2020 House," "FY2020 Senate," and "FY2020 Enacted" will be added, if available, as each action is completed.

[a]Amounts for the National Institute of Diabetes and Digestive and Kidney Diseases (NIDDK) do not include mandatory funding for type 1 diabetes research (see note h).

[b]Amounts for National Institute of General Medical Sciences (NIGMS) do not include funds from PHS Evaluation Set-Aside (§241 of the PHS Act). Though the budget request provides an increase to NIGMS through discretionary LHHS budget authority, the total amount for NIGMS with the PHS evaluation transfer included is less than FY2019 enacted levels.

[c]Amounts for National Institute of Environmental Health Sciences (NIEHS) do not include Interior/Environment Appropriations amount for Superfund research (see note g).

[d]Includes $12.6 million for the Gabriella Miller Kids First Research Act.

[e]The amount shown for the NIH Innovation Account in each column represents only a portion ($196 million for FY2019, $157 million for the FY2020 request) of the total appropriation to the account ($711 million for FY2020, $492 million for the FY2020 request). The remaining funds for this account are incorporated, where applicable, into the totals for other ICs. For the FY2020 request, this includes $195 to NCI for cancer research and $70 million to each of NINDS and NIMH for the BRAIN Initiative. (Continued ...)

[f]Amount for NIRSQ does not include the estimated $112.5 million in mandatory funding transfers from the Patient-Centered Outcomes Research Trust Fund (PCORTF) in FY2019, which were provided outside of the annual appropriations process. Mandatory funding for the PCORTF is scheduled to expire at the end of FY2019. The President's FY2020 budget did not request any new funding for the PCORTF.

[g]This is a separate account in the Interior/Environment appropriations for National Institute of Environmental Health Sciences (NIEHS) research activities related to Superfund. The FY2019 enacted amount is from Title III from the Consolidated Appropriations Act, 2019 (P.L. 116-6).

[h]Mandatory funds are available to NIDDK for type 1 diabetes research under PHS Act Section 330B (provided by §50902 P.L. 115-123 for FY2018 and FY2019). These funds are specified at $150 million annually. Under current law, no new funds will be available after September 30, 2019. The FY2020 request represents proposed funding.

NIH plans to continue to support biomedical innovations using new technologies, including for diagnosis, monitoring, and treatment. An example includes a smartphone-based system for people with diabetes to monitor blood glucose levels. NIH also aims to accelerate scientific discovery through new data science methods. In June 2018, NIH released a

Strategic Plan for Data Science, with an agency-wide plan for increasing and improving its use of large biomedical datasets. In addition, NIH plans to convene a new working group on artificial intelligence, machine learning, and biomedical research.[51]

Along with the above priorities, the President's budget request identifies ongoing support related to precision medicine, universal flu vaccine, and NIH buildings and facilities.[52]

DEPARTMENT OF ENERGY[53]

The Department of Energy (DOE) was established in 1977 by the Department of Energy Organization Act (P.L. 95-91), which combined energy-related programs from a variety of agencies with defense-related nuclear programs that dated back to the Manhattan Project. Today, DOE conducts basic scientific research in fields ranging from nuclear physics to the biological and environmental sciences; basic and applied R&D relating to energy production and use; and R&D on nuclear weapons, nuclear nonproliferation, and defense nuclear reactors. The department has a system of 17 national laboratories around the country, mostly operated by contractors, that together account for about 40% of all DOE expenditures.

The Administration's FY2020 budget request for DOE includes about \$12.783 billion for R&D and related activities, including programs in three broad categories: science, national security, and energy. This request is 18.6% less than the enacted FY2019 amount of \$15.712 billion. (See Table 10 for details).

The request for the DOE Office of Science is \$5.546 billion, a decrease of 15.8% from the FY2019 appropriation of \$6.585 billion. Funding would decrease for each of the office's six major research programs. In Basic Energy Sciences, almost half the proposed decrease would result from the approaching end of construction on the Linac Coherent Light Source II (no

[51] Ibid, p. 57, https://www.hhs.gov/sites/default/files/fy-2020-budget-in-brief.pdf.

[52] Ibid, p. 56.

[53] This section was written by Daniel Morgan, Specialist in Science and Technology Policy, CRS Resources, Science, and Industry Division.

funding requested for FY2020, down from $129 million in FY2019). Funding for Biological and Environmental Research would decrease by $211 million (29.9%), with reductions concentrated in the Earth and Environmental Systems Sciences subprogram. In Advanced Scientific Computing Research, the Office of Science Exascale Computing Project would receive $189 million, down 18.9% from $233 million in FY2019.[54] Funding for Fusion Energy Sciences would decrease by $161 million (28.6%), including a $25 million (18.9%) decrease in the U.S. contribution to construction of the International Thermonuclear Experimental Reactor (ITER), a fusion energy demonstration and research facility in France.

The request for DOE national security R&D is $4.925 billion, an increase of 11.8% from $4.406 billion in FY2019. Funding for Weapons Activities RDT&E would increase 37.9%, including an increase of $123 million for Advanced Simulation and Computing and an increase of $95 million (190.3%) for Enhanced Capabilities for Subcritical Experiments. In Defense Nuclear Nonproliferation R&D, reactor conversion activities would transfer to a non-R&D account; excluding this accounting change, funding for the remaining program would increase by 3.8%. Funding for the Naval Reactors program would decrease by 7.8% overall, with increases for operations, infrastructure, and technology development offset by previously planned decreases in funding for construction and two major multiyear projects.

The request for DOE energy R&D is $2.313 billion, a decrease of 51.0% from $4.721 billion in FY2019. Many of the proposed reductions in this category are similar to the Administration's FY2019 budget proposals. Funding for energy efficiency and renewable energy R&D would decrease by 66.3%, with reductions in all major research areas and a shift in emphasis toward early-stage R&D rather than later-stage development and deployment. Funding for fossil energy R&D would decrease by 24.1%, with reductions focused particularly on coal carbon capture and storage ($69 million, down from $199 million in FY2019) and natural gas technologies ($11 million, down from $51 million in FY2019).

---

[54] This project is part, but not all, of the DOE-wide Exascale Computing Initiative.

Table 10. Department of Energy R&D and Related Activities (budget authority, in millions of dollars)

	FY2019 Enacted	FY2020 Request	FY2020 House	FY2020 Senate	FY2020 Enacted
Science	$6,585	$5,546			
Basic Energy Sciences	2,166	1,858			
High Energy Physics	980	768			
Biological and Environmental Research	705	494			
Nuclear Physics	690	625			
Advanced Scientific Computing Research	936	921			
Fusion Energy Sciences	564	403			
Other	545	477			
National Security	4,406	4,925			
Weapons Activities RDT&E	2,014	2,778			
Naval Reactors	1,789	1,648			
Defense Nuclear Nonproliferation R&D	576	495			
Defense Environmental Cleanup Technol. Devel.	28	3			
Energy	4,721	2,313			
Energy Efficiency and Renewable Energya	2,067	696			
Fossil Energy R&D	740	562			
Nuclear Energy	1,326	824			
Electricity Delivery R&D	132	156			
Cybersec., En. Security, & Emerg. Response R&D	90	75			
Advanced Research Projects Agency-Energy	366	0			
DOE, Total	15,712	12,783			

Source: FY2019 enacted from P.L. 115-244 and H.Rept. 115-929. FY2020 request from DOE FY2020 congressional budget justification, https://www.energy.gov/cfo/downloads/fy-2020-budget-justification. Notes: Totals may differ from the sum of the components due to rounding. Figures for the columns headed "FY2020 House," "FY2020 Senate," and "FY2020 Enacted" will be added, if available, as each action is completed. [a]. Excluding Weatherization and Intergovernmental Activities.

The request for fuel cycle R&D is $90 million (down from $264 million in FY2019), and nuclear energy would decrease by 37.9%, with no funding requested for the Integrated University Program ($5 million in FY2019) or the Supercritical Transformational Electric Power (STEP) R&D initiative ($5 million in FY2019). The Advanced Research Projects Agency-Energy (ARPA-E), which is intended to advance high-impact energy technologies that have too much technical and financial uncertainty to attract near-term private-sector investment, would be terminated.

NATIONAL AERONAUTICS AND SPACE ADMINISTRATION[55]

The National Aeronautics and Space Administration (NASA) was created in 1958 by the National Aeronautics and Space Act (P.L. 85-568) to conduct civilian space and aeronautics activities. NASA has research programs in planetary science, Earth science, heliophysics, astrophysics, and aeronautics, as well as development programs for future human spacecraft and for multipurpose space technology such as advanced propulsion systems. In addition, NASA operates the International Space Station (ISS) as a facility for R&D and other purposes.

The Administration is requesting about $17.845 billion for NASA R&D in FY2020. This is 0.1% less than FY2019 funding of about $17.865 billion. For a breakdown of these amounts, see Table 11. NASA R&D funding comes through five accounts: Science; Aeronautics; Space Technology (called Exploration Technology in the Administration's budget request); Exploration (Deep Space Exploration Systems in the request); and the ISS, Commercial Crew, and Commercial Low Earth Orbit (LEO) Development portions of Space Operations (called LEO and Spaceflight Operations in the request).

The FY2020 request for Science is $6.304 billion, a decrease of 8.7% from FY2019. Within this total, funding for Earth Science would decrease by $151 million (7.8%); funding for Planetary Science would decrease by $136 million (4.9%); and funding for Astrophysics would decrease by $347 million (29.1%). The request for Earth Science includes no funding for the Pre-Aerosol, Clouds, and Ocean Ecosystem (PACE) mission or the Climate Absolute Radiance and Refractivity Observatory (CLARREO) Pathfinder mission. PACE and CLARREO Pathfinder were also proposed for termination in the FY2018 and FY2019 budgets but were funded by

[55] This section was written by Daniel Morgan, Specialist in Science and Technology Policy, CRS Resources, Science, and Industry Division.

Congress. The request for Planetary Science includes $593 million for a mission to Jupiter's moon Europa, but in contrast to prior congressional direction, the mission would launch on a commercial rocket and would not include a lander. The Planetary Science request also includes $210 million for the Lunar Discovery and Exploration program, initiated in FY2019, to fund public-private partnerships for research using commercial lunar landers. The request for Astrophysics does not include funding for the Wide Field Infrared Space Telescope (WFIRST, $312 million in FY2019). The proposed increase of $48 million for the James Webb Space Telescope (JWST) would provide $155 million more for JWST in FY2020 than was projected in the FY2019 budget; this change reflects previously announced cost increases and schedule delays.

The FY2020 request for Aeronautics is $667 million, a decrease of 8.0% from FY2019. The request includes $104 million for the Low Boom Flight Demonstrator program, intended to demonstrate quiet supersonic flight.

The FY2020 request for Exploration Technology (currently Space Technology) is $1.014 billion, an increase of 9.4% from FY2019. The request proposes $119 million for a mostly new Lunar Surface Innovation Initiative. It proposes $45 million for a restructured In-Space Robotic Servicing program, down from $180 million for the RESTORE-L robotic servicing mission in FY2019.

The FY2020 request for Deep Space Exploration Systems (currently Exploration) is $5.022 billion, a decrease of 0.6% from FY2019. This account funds development of the Orion Multipurpose Crew Vehicle and the Space Launch System (SLS) heavy-lift rocket, the capsule and launch vehicle mandated by the NASA Authorization Act of 2010 for future human exploration beyond Earth orbit. The first test flight of SLS carrying Orion but no crew (known as EM-1) is now expected no earlier than June 2020. The first flight of Orion and the SLS with a crew on board (known as EM-2) is expected by April 2023.

Table 11. National Aeronautics and Space Administration R&D (budget authority, in millions of dollars)

	FY2019 Enacted	FY2020 Request	FY2020 House	FY2020 Senate	FY2020 Enacted
Science	6,906	6,304			
Earth Science	1,931	1,780			
Planetary Science	2,759	2,622			
Astrophysics	1,192	845			
James Webb Space Telescope	305	353			
Heliophysics	720	705			
Aeronautics	725	667			
Space Tech./Exploration Tech.	927	1,014			
Exploration/Deep Sp. Exp. Sys.	5,051	5,022			
Exploration Systems Development	4,093	3,442			
Exploration R&D	958	1,580			
Space Ops./LEO and Spaceflight Ops.a	~1,678a	1,710a			
International Space Station	n/s	1,458			
Commercial Crew	173	102			
Commercial LEO Development	40	150			
Subtotal R&D	15,287	14,567			
Non-R&D Programsb	3,110	2,617			
Safety, Security, and Mission Services	2,755	3,085			
Associated with R&Dc	2,289	2,619			
Construction & Environmental C&R	348	600			
Associated with R&Dc	289	510			
NASA, Total (R&D)	17,865	17,845			
NASA, Total	21,500	21,019			

Sources: FY2019 enacted from P.L. 116-6 and H.Rept. 116-9. FY2020 request from NASA FY2020 congressional budget justification, http://www.nasa.gov/news/budget/.

Notes: Totals may differ from the sum of the components due to rounding. LEO = Low Earth Orbit. C&R = Compliance and Remediation. Figures for the columns headed "FY2020 House," "FY2020 Senate," and "FY2020 Enacted" will be added, if available, as each action is completed.

[a]Excluding non-R&D activities: Space and Flight Support and Space Transportation other than Commercial Crew. FY2019 amount includes CRS estimate of $1.465 billion for International Space Station for the purpose of calculating a total.

[b]Non-R&D activities in Space Operations / LEO and Spaceflight Operations (see note a); STEM Engagement (formerly Education); and Inspector General.

[c]CRS estimates the allocation between R&D and non-R&D in proportion to the underlying program amounts in order to allow calculation of a total for R&D. The Safety, Security, and Mission Services account and the Construction and Environmental Compliance and Remediation account consist mostly of indirect costs for other programs, assessed in proportion to their direct costs.

Funding for Orion, the SLS, and related ground systems (collectively known as Exploration Systems Development) would decrease by $651 million (15.9%). The account also funds Exploration R&D, which would increase by $622 million (64.9%). The request for Exploration R&D includes $821 million for a platform in lunar orbit (known as the Gateway) to serve as a test bed for deep space human exploration capabilities. In the LEO and Spaceflight Operations account (currently Space Operations), the request for Commercial Crew is $102 million, a decrease of 41.1% from FY2019; the request for the ISS is $1.458 billion;[56] and the request for Commercial LEO Development, a new program in FY2019, is $150 million (an increase of 275.0%). The reduction in Commercial Crew funding reflects the expected transition of commercial crew activities from development to operations: the first post-certification crewed flight to the ISS is planned for late FY2019. The Commercial LEO Development program is intended to stimulate a commercial space economy in low Earth orbit, including the commercial provision of NASA's requirements for research and technology demonstration after the proposed end of direct ISS funding in 2025.

NATIONAL SCIENCE FOUNDATION[57]

The National Science Foundation supports basic research and education in the nonmedical sciences and engineering. Congress established the foundation as an independent federal agency in 1950 and directed it to "promote the progress of science; to advance the national health, prosperity, and welfare; to secure the national defense; and for other purposes."[58] The NSF is a primary source of federal support for U.S. university research, especially in mathematics and computer science. It is also responsible for significant shares of the federal science, technology,

---

[56] The FY2019 appropriations act and conference report did not specify an FY2019 amount for the ISS.

[57] This section was written by Laurie Harris, Analyst in Science and Technology Policy, CRS Resources, Science, and Industry Division.

[58] The National Science Foundation Act of 1950 (P.L. 81-507).

engineering, and mathematics (STEM) education program portfolio and federal STEM student aid and support.

NSF has six appropriations accounts: Research and Related Activities (RRA, the main research account), Education and Human Resources (EHR, the main education account), Major Research Equipment and Facilities Construction (MREFC), Agency Operations and Award Management (AOAM), the National Science Board (NSB), and the Office of Inspector General (OIG). Appropriations are generally provided at the account level, while program-specific direction may be included in appropriations acts, or accompanying conference reports or explanatory statements.

Because final FY2019 funding was not available at the time the FY2020 budget request was prepared, requested R&D funding is compared to the FY2018 actual funding. FY2019 funding levels, enacted February 15, 2019, are included for reference.[59] These amounts are available only at the account level; FY2019 R&D breakouts and subaccount funding amounts are not yet available for comparison.

Funding for R&D is included in the RRA, EHR, and MREFC accounts. (The RRA and EHR accounts also include non-R&D funding). Together, these three accounts comprise 95% of the total requested funding for NSF. Actual R&D obligations for each account are known after NSF allocates funding appropriations to specific activities and reports those figures.[60] The budget request specifies R&D funding for the conduct of research, including basic and applied research, and for physical assets, including R&D facilities and major equipment. Funding amounts for FY2018 actual and FY2020 requested levels are reported by account, including amounts for R&D conduct and physical assets where applicable, in Table 12.

---

[59] The Consolidated Appropriations Act, 2019 (P.L. 116-6); and Explanatory Statement, Consolidated Appropriations Act, 2019, Division C (Commerce, Justice, Science, and Related Agencies Appropriations Act, 2019), *Congressional Record*, vol. 165, no. 28—Book II (February 13, 2019), p. H1813.

[60] R&D actual (FY2018) and requested (FY2020) amounts are reported in the "Quantitative Data Tables" section of the FY2020 *NSF Budget Request to Congress*, March 18, 2019, pp. QDT-1–QDT-7.

Table 12. National Science Foundation Funding (budget authority, in millions of dollars)

Account	FY2018 Actual	FY2019 Enacted	FY2020 Request	FY2020 House	FY2020 Senate	FY2020 Enacted
Research and Related Activities (RRA)	6,380.4	6,520.0	5,663.0			
R&D, RRA Total	5,714.1	n/a	5,079.7			
Conduct of R&D	5,391.8	n/a	4,797.3			
R&D Facilities and Major Equipment	322.3	n/a	282.4			
Education and Human Resources	903.9	910.0	823.5			
(EHR)						
R&D, EHR Total	457.7	n/a	420.0			
Conduct of R&D	457.7	n/a	420.0			
R&D Facilities and Major Equipment	0.0	n/a	0.0			
Major Research Equipment and	186.3	295.7	223.2			
Facilities Construction (MREFC)						
R&D, MREFC Total	186.3	n/a	223.2			
Conduct of R&D	0.0	n/a	0.0			
R&D Facilities and Major Equipment	186.3	n/a	223.2			
Agency Operations and Award	328.5	329.5	336.9			
Management (AOAM)a						
National Science Board (NSB)a	4.3	4.4	4.1			
Office of the Inspector General (OIG)a	15.1	15.4	15.4			
NSF, Total Discretionaryb	7,818.4	8,075.0	7,066.0			
R&D, NSF Total	6,358.0	n/a	5,722.9			
Total, Conduct of R&D	5,849.5	n/a	5,217.3			
Total, R&D Facilities & Major Equipment	508.6	n/a	505.6			

Source: Data in the columns titled, "FY2018 Actual" and "FY2020 Request" are from the FY2020 *NSF Budget Request to Congress*. Data in the "FY2019 Enacted" column are from P.L. 116-6.

Notes: Appropriations accounts are in bold. NSF total may differ from the sum of the accounts due to rounding. Nonbold R&D funding amounts are a subset of funding for the specified accounts. The term "n/a" means not available. Figures for the columns headed "FY2020 House," "FY2020 Senate," and "FY2020 Enacted" will be added, if available, as each action is completed.

[a]The AOAM, NSB, and OIG accounts have no reported R&D funding.

[b]In addition to discretionary funding, NSF reports mandatory funding from H-1B visa and donation sources, which are not included in this total.

Overall

The Administration is requesting $7.07 billion for the NSF in FY2020, $1.01 billion (12.5%) less than the FY2019 enacted amount, and $752 million (9.6%) less than the FY2018 actual amount. The request would decrease budget authority in three accounts relative to the FY2018 enacted level: RRA by $717.4 million (11.2%), EHR by $80.4 million (8.9%), and NSB by $200,000 (4.7%). The request would increase budget authority for the MREFC account by $36.9 million (19.8%) and provide slight increases to the AOAM (2.6%, $8.4 million) and OIG (1.7%, $260,000) accounts. Overall, NSF estimates that, under the FY2020 request, agency-wide funding rates (i.e., the percentage of submitted proposals that are successfully awarded funding) would decrease slightly from 24% to 23%, with 1,317 fewer new competitive awards, compared to FY2018.

As a proportion of NSF's total funding, R&D activities account for approximately 81%. For FY2020, $5.72 billion is requested for R&D activities, a 10% decrease from FY2018 actual funding for R&D of $6.36 billion. The total request includes $5.22 billion (91%) for the conduct of R&D, and $506 million (9%) for R&D facilities and major equipment. Of funding requested for the conduct of R&D, 87% is requested for basic research, and 13% for applied research. Overall funding for R&D facilities and major equipment supports not only the construction and acquisition phases, funded through MREFC ($223 million requested), but also the planning, design, and postconstruction operations and maintenance, funded through RRA ($282 million requested).

Research

The Administration seeks $5.66 billion for RRA in FY2020, an $857 million (13.1%) decrease compared to the FY2019 enacted funding, and a $717 million (11.2%) decrease compared to FY2018 actual funding. Compared to the FY2018 actual levels, the FY2020 request includes decreases for 8 of the 10 RRA subaccounts. The largest percentage decrease would go to the Office of Polar Programs (19.6%, $98.3 million decrease). The largest percentage increase would go to the U.S. Arctic Research Commission account (6.3%, $90,000 increase). The FY2020 request also includes $151 million for the RRA Established Program to

Stimulate Competitive Research (EPSCoR) program, a $19.4 million (11.3%) decrease compared to FY2018 actual funding.

Within the RRA account, the FY2020 request includes $5.08 billion for R&D, a decrease of $634 million (11.1%) compared to the FY2018 actual amount. Of this amount, the majority ($4.80 billion, 94%) is requested for the conduct of research, including $4.38 billion for basic research and $417 million for applied research.

Education

The FY2020 request for the EHR account is $86.5 million (9.5%) less than the FY2019 enacted amount and $80.4 million (8.9%) less than the FY2018 actual level. By program division, the Division of Human Resource Development would receive an increase of $15.6 million (9.6%) over the FY2018 actual level. The divisions of research on learning in formal and informal settings, graduate education, and undergraduate education would receive decreases of 20.4% ($182 million requested), 5.5% ($244 million requested), and 13.8% ($219 million requested), respectively.

EHR programs of particular interest to congressional policymakers include the Graduate Research Fellowship Program (GRFP) and National Research Traineeship (NRT) programs. The FY2020 request for GRFP is $257 million, a reduction of $27.9 million (9.8%) from the FY2018 actual level. The FY2020 request for NRT is $49.5 million, a $4.3 million (8.0%) decrease from FY2018.

Within EHR, requested funding for R&D is $420 million, which is $37.7 million (8.2%) less than the FY2018 actual funding amount and accounts for approximately 7.3% of the agency's total R&D request. All of the requested funding would support the conduct of R&D, including $150 million for basic research and $270 million for applied research.

Construction

The MREFC account supports large construction projects and scientific instruments, with all of the funding supporting R&D facilities. The Administration is seeking $223 million for MREFC in FY2020, $36.9 million (19.8%) more than the FY2018 enacted amount, and $72.5 million (24.5%) less than the FY2018 actual amount.

Requested MREFC funding would support continued construction of the Large Synoptic Survey Telescope (LSST, $46.3 million requested, 5.1% decrease from FY2019 enacted) and Antarctic Infrastructure Modernization for Science (AIMS, $97.9 million requested, 5.6% decrease from FY2019 enacted). The request includes $33.0 million for upgrades to the Large Hadron Collider in Switzerland, which would represent the first year of a five-year project. Additionally, $45.0 million is requested for Mid-scale Research Infrastructure projects in the $20 million to $70 million range; this is a new funding line-item in the MREFC account meant to manage support for upgrades to major facilities and stand-alone projects in this range as a portfolio.

Other Initiatives

The FY2019 NSF budget request includes funding for three multiagency initiatives. This funding is included in multiple NSF appropriations accounts, and R&D amounts are not separately provided. The National Nanotechnology Initiative would receive $389 million, $179 million (31.4%) less than in FY2018. The Networking and Information Technology Research and Development program would receive $1.20 billion, a decrease of $98.2 million (7.6%). The U.S. Global Change Research Program would receive $224 million, $30.0 million (11.8%) less than in FY2018.[61]

[61] For additional information on these initiatives, see "Multiagency R&D Initiatives."

DEPARTMENT OF AGRICULTURE[62]

The U.S. Department of Agriculture (USDA) was created in 1862, in part to support agricultural research in an expanding, agriculturally dependent country. Today, USDA conducts intramural research at federal facilities with federally employed scientists and supports external research at universities and other facilities through competitive grants and formula-based funding. The breadth of contemporary USDA research spans traditional agricultural production techniques, organic and sustainable agriculture, bioenergy, nutrition needs and food composition, food safety, animal and plant health, pest and disease management, economic decisionmaking, and other social sciences affecting consumers, farmers, and rural communities.

Four agencies carry out USDA's intramural research and education activities, grouped together into the Research, Education, and Economics (REE) mission area. The agencies involved are the Agricultural Research Service (ARS), National Institute of Food and Agriculture (NIFA), National Agricultural Statistics Service (NASS), and Economic Research Service (ERS).

The FY2019 enacted appropriation (P.L. 116-6) provides a total of $3,424.1 million in discretionary spending for the four agencies. The Administration is requesting a total of $2,868.7 million for the four agencies in FY2020, a 16.2% reduction ($555.4 million). Most of that year-over-year reduction (78%) is attributable to the reduced request for ARS salaries and expenses and the ARS buildings and facilities account (see Table 13). The remainder of the year-over-year reduction comes from decreases in certain research and education, extension, and integrated activities in NIFA, as well as in NASS and ERS. In addition to discretionary appropriations, agricultural research is funded by state matching contributions and private donations or grants, as well as certain

[62] This section was written by Tadlock Cowan, Analyst in Natural Resources and Rural Development, CRS Resources, Science, and Industry Division.

mandatory funding authorized by the farm bill.[63] USDA's FY2019 enacted discretionary appropriations and the Administration's FY2020 request for the four research agencies are discussed below.

Agricultural Research Service

The Agricultural Research Service is USDA's in-house basic and applied research agency. It operates approximately 90 laboratories nationwide with about 6,600 employees. ARS laboratories focus on efficient food and fiber production, development of new products and uses for agricultural commodities, development of effective controls for pest management, and support of USDA regulatory and technical assistance programs. ARS also operates the National Agricultural Library, one of the department's primary information repositories for food, agriculture, and natural resource sciences.

For FY2019, P.L. 116-6 provides $1,303.3 billion for ARS salaries and expenses and $381.2 million for buildings and facilities. The Administration is requesting $1,203.5 billion for ARS salaries and expenses for FY2020, a decrease of $99.8 million (7.6%) from the FY2019 appropriation. For FY2020, the request for the buildings and facilities account is $50.0 million (Table 13).

ARS will assume ownership of the National Bio and Agro-Defense Facility (NBAF) in FY2019 from the Department of Homeland Security (DHS). The FY2019 enacted bill provides $10.6 million to address one-time costs associated with the transfer of the science program from the Plum Island Animal Disease Center to NBAF, and $42.0 million to address stand-up activities and other initial costs to operate and maintain the new facility. NBAF is expected to be fully operational by December 31, 2022. From the total salaries and expenses appropriation for FY2020, the Administration is requesting $13.1 million for NBAF. The FY2019 enacted bill provides an additional $5.0 million for ARS to increase

[63] For additional information, see CRS Report R45230, *Agriculture and Related Agencies: FY2019 Appropriations*, by Jim Monke.

research efforts on foreign animal diseases, and an additional $2.0 million to expand research on resilient dryland farming.

Table 13. U.S. Department of Agriculture R&D (budget authority, in millions of dollars)

Agency or Major Program	FY2019 Enacted P.L. 116-6	FY2020 Request	FY2020 House	FY2020 Senate	FY2020 Enacted
Agricultural Research Service					
Salaries and Expenses	1,303.3	1,203.5			
Buildings and Facilities	381.2	50.0			
Other	7.0				
Subtotal, ARS	1,691.5	1,253.5			
National Institute of Food and Agriculture (NIFA)					
Research and Education					
AFRI (competitive grants)	415.0	500.0			
Hatch Act (1862 institutions)	259.0	243.2			
Evans-Allen (1890 institutions)	58.0	53.8			
McIntire-Stennis (forestry)	36.0	28.9			
Other	159.6	148.8			
Subtotal	927.6	974.7			
Extension					
Smith-Lever (b) and (c)	315.0	299.4			
Smith-Lever (d)	86.6	58.1			
1890 Extension Activities	48.6	47.3			
1994 Extension Activities	6.4	4.4			
Other	49.1	6.1			
Subtotal	505.7	415.3			
Integrated Activities	38.0	1.7			
Subtotal, NIFA	1,471.3	1,391.7			
National Agricultural Statistics Service (NASS)	174.5	163.0			
Economic Research Service (ERS)	86.8	60.5			
Total, USDA Research, Education, and Economics Mission Area	3,424.1	2,868.7			

Source: CRS, compiled from P.L. 116-6, Conf. Rept. 116-9; *Omnibus Appropriation Explanatory Statement, Division A*; *2020 USDA Budget Justification Notes.* Figures for the columns headed "FY2020 House," "FY2020 Senate," and "FY2020 Enacted" will be added, if available, as each action is completed.

FY2019 conference report language (H.Rept. 116-9) criticized ARS for not reporting a single specific negative finding by Animal and Plant Health Inspection Service (APHIS) inspections of ARS research facilities that use animals as research subjects. The report noted that numerous violations had been found involving death and serious health issues of animal subjects, and directed ARS to submit a report within 60 days of enactment covering all violations found by APHIS and the actions taken to prevent them from recurring.

P.L. 116-9 does not support the Administration's request to terminate or redirect various ARS research programs, or the closure of ARS research locations.

National Institute of Food and Agriculture

The National Institute of Food and Agriculture (NIFA) provides federal funding for research, education, and extension projects conducted in partnership with State Agricultural Experiment Stations, the State Cooperative Extension System, land grant universities, colleges, and other research and education institutions, as well as individual researchers. These partnerships include the 1862 land-grant institutions, 1890 historically black colleges and universities (HBCUs) established by the Morrill Acts, the 1994 tribal land-grant colleges, and Hispanic-serving institutions.[64] Federal funds enhance research capacity at universities and institutions through statutory formula funding, competitive awards, and grants.[65]

For FY2019, P.L. 116-6 provides $1,471.3 billion in discretionary spending for NIFA activities. The Administration's FY2020 request for NIFA is $1,391.7 billion, a reduction of $79.6 million (5.4%) from FY2019 (Table 13). The enacted bill provides $259.0 million to support Hatch Act formula funding for 1862 land grant university research and education activities. For FY2020, the Administration is requesting $243.2

---

[64] The numbers 1862, 1890, and 1994 in this context refer to the years that laws were enacted creating these classifications of colleges and universities, not to the number of institutions.

[65] The National Agricultural Research, Extension, and Teaching Policy Act of 1977 designated USDA as the lead federal agency for higher education in the food and agricultural sciences.

million for Hatch Act funding, a 6.1% reduction. For Evans-Allen formula funding to the 19 HBCUs, the FY2019 bill provides \$58.0 million for research and \$19.3 million for education grants. The Administration requests \$53.8 million in Evans-Allen funding for FY2020 (7.2% reduction from FY2019), and \$18.7 million for education grants. For research grants to the 1994 Tribal institutions, and for education grants to Alaska Native and Native Hawaiian-Serving institutions, the FY2019 appropriation provides \$3.8 and \$3.2 million, respectively. For FY2020, the Administration requests \$3.4 million for the 1994 Tribal institutions, and \$0 for education grants to the Alaska Native and Native Hawaiian-Serving institutions.

For McIntire-Stennis cooperative forestry research support, P.L. 116-6 provides \$36.0 million for FY2019. The Administration is requesting \$28.9 million for FY2020, approximately 20% less than FY2019. The FY2019 appropriation also provides \$37.0 million for the Sustainable Agriculture Research and Education program. The Administration requests a reduction of \$18.0 million (48.6%) for the program in FY2020.

The FY2019 enacted bill provides \$415.0 million for the Agriculture and Food Research Initiative (AFRI)—USDA's flagship competitive research grants program. The Administration is requesting \$500.0 million for the program in FY2020, a 20.5% increase over FY2019. This budget item currently represents about 30% of the total NIFA discretionary budget.

For Cooperative Extension support at 1862 land grant universities under the Smith-Lever Act, Sections (b) and (c) formula funding for FY2019, the enacted appropriation provides a total of \$315.0 million for these extension activities. The Administration requests \$299.4 million for these programs in FY2020. The Smith-Lever Sections (b) and (c) programs include extension services at the HBCUs and the 1994 Tribal colleges, faculty improvement grants to HBCUs, and women and minorities in STEM fields, among other programs.

P.L. 116-6 provides \$86.6 million for Smith-Lever 3(d) activities, including food and nutrition education, new technologies for agricultural extension, and children, youth, and families at risk. For FY2020, the

Administration is requesting $58.1 million for Smith-Lever Section 3(d) funding, $55.1 million of which would support the Expanded Food and Nutrition Education Program, and $3.0 million would support Federally-Recognized Tribes Extension Program for programs on American Indian Reservations and Tribal jurisdictions. The Administration is requesting $0 funding in FY2020 for other Smith-Lever Section 3(d) programs.

National Agricultural Statistics Service

The National Agricultural Statistics Service conducts the quinquennial Census of Agriculture and provides official statistics on agricultural production and indicators of the economic and environmental status of the farm sector.

For FY2019, P.L. 116-6 provides $174.5 million to NASS, of which up to $45.3 million is reserved to support the 2017 Census of Agriculture. The enacted bill also provides $600,000 for the Geospatial Improvement Initiative and an increase of $500,000 for the Floriculture Crops Report. The Administration is requesting $163.0 million for NASS in FY2020, and up to $45.3 million to support the 2017 Census. Results of the 2017 Census of Agriculture were released on April 11, 2019.

Economic Research Service

The Economic Research Service supports economic and social science analysis about agriculture, rural development, food, commodity markets, and the environment. It also collects and disseminates data concerning USDA programs and policies. ERS is one of 13 "principal statistical agencies" of the Federal Statistical System of the United States.[66]

[66] In August 2018, the Secretary announced plans to relocate ERS outside the Washington D.C. region. Congress has asked for a cost-benefit analysis of the proposed move, and USDA's Inspector General is conducting a review of the authorities of the department to effect a relocation. For further information on the proposed relocation of ERS, see CRS In Focus IF11166, *Proposed Relocation/Realignment of USDA's ERS and NIFA*, by Tadlock Cowan.

For FY2019, P.L. 116-6 provides $86.8 million for ERS activities. The Administration has requested $60.5 million for ERS in FY2020, a 30.3% decrease.

DEPARTMENT OF COMMERCE

Two agencies of the Department of Commerce have major R&D programs: the National Institute of Standards and Technology (NIST) and the National Oceanic and Atmospheric Administration (NOAA).

National Institute of Standards and Technology[67]

The mission of the National Institute of Standards and Technology is "to promote U.S. innovation and industrial competitiveness by advancing measurement science, standards, and technology in ways that enhance economic security and improve our quality of life."[68] NIST research provides measurement, calibration, and quality assurance methods and techniques that support U.S. commerce, technological progress, product reliability, manufacturing processes, and public safety. NIST's responsibilities include the development, maintenance, and custodial retention of the national standards of measurement; providing the means and methods for making measurements consistent with those standards; and ensuring the compatibility of U.S. national measurement standards with those of other nations.[69]

The President is requesting $686.8 million for NIST in FY2020, a decrease of $298.7 million (30.3%) from the FY2019 enacted appropriation of $985.5 million. (See Table 14). NIST discretionary funding is provided through three accounts: Scientific and Technical

[67] This section was written by John F. Sargent Jr., Specialist in Science and Technology Policy, CRS Resources, Science, and Industry Division.
[68] NIST website, "General Information," http://nist.gov/public_affairs/general_information.cfm.
[69] 15 U.S.C. §272.

Research and Services (STRS), Industrial Technology Services (ITS), and Construction of Research Facilities (CRF).

The President's FY2020 request includes $611.7 million for R&D, standards coordination, and related services in the STRS account, a decrease of $112.8 million (15.6%) from the FY2019 level.[70]

Table 14. National Institute of Standards and Technology Funding (budget authority, in millions of dollars)

	FY2019 Enacted	FY2020 Request	FY2020 House	FY2020 Senate	FY2020 Enacted
Scientific and Technical Research and Services	$724.5	$611.7			
Industrial Technology Services	155.0	15.2			
Manufacturing Extension Partnership	140.0	0.0			
Manufacturing USA / National Network for Manufacturing Innovation	15.0	15.2			
Construction of Research Facilities	106.0	59.9			
NIST, Totala	985.5	686.8			

Source: U.S. Department of Commerce, National Institute of Standards and Technology, National Institute of Standards and Technology/National Technical Information Service, Fiscal Year 2020 Budget Submission to Congress, p. NIST - 5, https://www.commerce.gov/sites/default/files/2019-03/ fy2020_nist_congressional_budget_justification.pdf.

Notes: Figures for the columns headed "FY2020 House," "FY2020 Senate," and "FY2020 Enacted" will be added, if available, as each action is completed.

[a] The NIST FY2020 budget justification notes that "The budgetary resources from offsetting collections for the NIST Public Safety Communications Research Fund will obligate over several fiscal years." (These funds are not included in the table figures). These funds were provided through the NIST Public Safety Communications Research Fund to help develop wireless technologies for public safety users, as part of the National Wireless Initiative included in the Middle Class Tax Relief and Job Creation Act of 2012 (P.L. 112- 96). This act provides mandatory funds for NIST from spectrum auction proceeds to help industry and public safety organizations conduct research and develop new standards, technologies, and applications to advance public safety communications in support of the initiative's efforts to build an interoperable nationwide broadband network for first responders. The act provided NIST a total of $300 million, though rescissions reduced this amount to $285 million.

[70] CRS analysis of data from U.S. Department of Commerce, National Institute of Standards and Technology, National Institute of Standards and Technology/National Technical Information Service, Fiscal Year 2020 Budget Submission to Congress, p. NIST-5, https://www.commerce.gov/sites/default/files/2019-03/ fy2020_nist_congressional_budget_justification.pdf.

The FY2020 request would provide $15.2 million for the ITS account, down $139.8 million (90.2%) from FY2019. Within the ITS account, the request would provide no funding for the Manufacturing Extension Partnership (MEP) program, a reduction of $140.0 million from FY2019; MEP centers in each state would be required to become entirely self-supporting. In his FY2019 request, President Trump proposed ending federal funding for MEP; in his FY2018 request, the President sought $6.0 million "for an orderly shutdown of the program." The request provides $15.2 million provided for Manufacturing USA (also referred to as the National Network for Manufacturing Innovation or NNMI), slightly higher than the FY2019 level of $15.0 million. Of these funds, approximately $10 million would be for continued support of the NISTsponsored National Institute for Innovation in Manufacturing Biopharmaceuticals (NIIMBL), with the balance (approximately $5 million) to be used for coordination of the Manufacturing USA network.[71]

The President is requesting $59.9 million for the NIST CRF account for FY2020, down $46.1 million (43.5%) from the FY2019 enacted level.[72]

National Oceanic and Atmospheric Administration[73]

The National Oceanic and Atmospheric Administration (NOAA) conducts scientific research in areas such as ecosystems, climate, global climate change, weather, and oceans; collects and provides data on the oceans and atmosphere; and manages coastal and marine organisms and environments. NOAA was created in 1970 by Reorganization Plan No. 4.[74] The reorganization was intended to unify elements of the nation's environmental programs and to provide a systematic approach for monitoring, analyzing, and protecting the environment.

[71] Ibid.

[72] Ibid.

[73] This section was written by Eva Lipiec, Analyst in Natural Resources Policy, CRS Resources, Science, and Industry Division.

[74] "Reorganization Plan No. 4 of 1970," 35 *Federal Register* 15627-15630, October 6, 1970.

NOAA's administrative structure is organized by six line offices that reflect its diverse mission: the National Ocean Service (NOS); National Marine Fisheries Service (NMFS); National Environmental Satellite, Data, and Information Service (NESDIS); National Weather Service (NWS); Office of Oceanic and Atmospheric Research (OAR); and the Office of Marine and Aviation Operations (OMAO). The line offices are supported by an additional office, Mission Support, which provides cross-cutting administrative functions related to planning, information technology, human resources, and infrastructure. Congress provides most of the discretionary funding for the line offices and Mission Support through two accounts: (1) Operations, Research, and Facilities, and (2) Procurement, Acquisition, and Construction.

In 2010, NOAA published its *Next Generation Strategic Plan*.[75] The strategic plan is organized into four categories of long-term goals including (1) climate adaptation and mitigation, (2) a weather-ready nation,[76] (3) healthy oceans, and (4) resilient coastal communities and economies.

The strategic plan also lists three groups of enterprise objectives related to (1) stakeholder engagement, (2) data and observations, and (3) integrated environmental modeling.[77] The strategic plan serves as a guide for NOAA's five-year R&D plan. The most recent five-year R&D plan was published in 2013, and includes R&D objectives to reach strategic plan goals and objectives and targets to track progress toward R&D objectives over time.[78]

---

[75] NOAA, *NOAA's Next-Generation Strategic Plan*, Silver Spring, MD, December 2010, at https://www.performance.noaa.gov/wp-content/uploads/NOAA_NGSP.pdf.

[76] According to NOAA, a weather-ready nation is envisioned as a society that is prepared for and responds to weather-related events.

[77] NOAA defines the enterprise objectives as "cross-cutting requirements for addressing NOAA's strategic goals as a whole." NOAA, *NOAA's Next-Generation Strategic Plan*, Silver Spring, MD, December 2010, p. 32.

[78] NOAA, *Research and Development at NOAA, Five-Year Research and Development Plan 2013-2017*, 2014, at http://nrc.noaa.gov/CouncilProducts/ResearchPlans/5YearRDPlan/NOAA5YRPHome/Preface/Purpose.aspx. NOAA's Research Council is charged with developing the five-year research and development plans. According to NOAA, "The NOAA Research Council is an internal body composed of senior scientific personnel from every line office in the agency who provide corporate oversight to ensure NOAA's research and development activities are of the highest quality, meet near- to long-term mission requirements and societal needs, take advantage of emerging scientific and technological opportunities, shape a forward-looking research agenda, and are accomplished in an

One of the main challenges identified in the NOAA R&D plan is the need to integrate the diverse perspectives and professional expertise required by the agency's mission. The plan states that "holistically understanding the earth system is not only understanding its individual components, but understanding and interpreting the way each of the components interact and behave as an integrated composite that is more than the sum of its parts."[79]

For FY2020, President Trump requested $651.1 million in R&D funding for NOAA, a decrease of $286.9 million (30.6%) below the FY2019 enacted level of $938.0 million. For FY2019, Congress enacted $540.3 million for research (57.6% of total R&D funding), $162.5 million for development (17.3%), and $235.2 million for R&D equipment (25.1%).[80] The enacted FY2019 total R&D amount was 17.0% of NOAA's total discretionary budget authority of $5.509 billion.[81] In FY2020, the President is requesting $352.3 million for research (54.1% of total R&D funding), $106.3 million for development (16.3%), and $192.6 million for R&D equipment (29.6%).[82] The President's request for total R&D is 14.1% of NOAA's total discretionary budget authority request of $4.622 billion.[83]

Table 15 provides R&D funding levels for FY2019 enacted and the Administration's FY2020 request for each NOAA office.[84] OAR accounts for the majority of R&D in most years. The President is requesting $335.1

efficient and cost-effective manner." NOAA website, "NOAA Research Council," at http://nrc.noaa.gov.

[79] NOAA, *Research and Development at NOAA, Five-Year Research and Development Plan 2013-2017*, 2014.

[80] Email correspondence between the NOAA Budget Office and CRS, April 8, 2019.

[81] NOAA, *Budget Estimates Fiscal Year 2020*, 2019. NOAA's requested budget uses several budgetary terms, such as direct obligations, budget authority, and appropriations. For more information, see CRS In Focus IF11185, *National Oceanic and Atmospheric Administration (NOAA): FY2020 Budget Request and Appropriations*, by Eva Lipiec.

[82] Email correspondence between the NOAA Budget Office and CRS, April 8, 2019.

[83] NOAA, *Budget Estimates Fiscal Year 2020*, 2019, at https://www.corporateservices .noaa.gov/nbo/fy20_bluebook/ NOAA-FY20-Congressional-Justification.pdf.

[84] Email correspondence between the NOAA Budget Office and CRS, April 8, 2019.

million for OAR R&D in FY2020, a decrease of $196.2 million (36.9%) below the FY2019 enacted funding level of $531.4 million.[85]

OAR conducts research in three major areas: (1) weather and air chemistry; (2) climate; and (3) oceans, coasts, and the Great Lakes. A significant portion of these efforts is implemented through NOAA laboratories and cooperative research institutes. NOAA supports 16 cooperative research institutes and 10 NOAA laboratories in OAR's three research areas. The President's FY2020 request would fund the cooperative institutes and laboratories at $169.6 million, $13.1 million (7.2%) less than the FY2019 enacted funding level of $182.8 million.[86]

Table 15. National Oceanic and Atmospheric Administration R&D (budget authority, in millions of dollars)

	FY2019 Enacted	FY2020 Request	FY2020 House	FY2020 Senate	FY2020 Enacted
National Ocean Service (NOS)	103.2	54.1			
National Marine Fisheries Service (NMFS)	60.1	55.4			
National Weather Service (NWS)	26.1	18.3			
National Environmental Satellite, Data, and Information Service (NESDIS)	31.0	27.8			
Office of Marine and Aviation Operations (OMAO)a	161.2	160.3			
Office of Oceanic and Atmospheric Research (OAR)	531.4	335.1			
Mission Support	25.0	0			
Total R&D	938.0	651.1			
NOAA, Total R&D and Non-R&D	5,509.0	4,621.7			

Source: Email correspondence between the NOAA Budget Office and CRS, April 8, 2019, and NOAA, *Budget Estimates Fiscal Year 2020*, 2019, at https://www.corporateservices.noaa.gov/nbo/fy20_bluebook/NOAA-FY20- Congressional-Justification.pdf.

Notes: Figures for the columns headed "FY2020 House," "FY2020 Senate," and "FY2020 Enacted" will be added, if available, as each action is completed. Totals may differ from the sum of the components due to rounding.

[a]All Office of Marine Aviation Operations funding is for equipment related to R&D.

[85] Email correspondence between the NOAA Budget Office and CRS, April 8, 2019. Totals may differ from the sum of the components due to rounding.

[86] NOAA, *Budget Estimates Fiscal Year 2020*, 2019. Totals may differ from the sum of the components due to rounding.

Among other R&D activities, the President's FY2019 request would also reduce funding to the National Sea Grant College Program. The National Sea Grant College Program is composed of 33 university-based state programs and supports scientific research and stakeholder engagement to identify and solve problems faced by coastal communities. The President's FY2020 request would terminate federal support of the National Sea Grant College Program and its related Marine Aquaculture Research program. In FY2019, Congress appropriated $68.0 million to the National Sea Grant College Program and $12.0 million to the Marine Aquaculture Research program.[87]

DEPARTMENT OF THE INTERIOR[88]

The Department of the Interior (DOI) was created to conserve and manage the nation's natural resources and cultural heritage, to provide scientific and other information about those resources, and to uphold "the nation's trust responsibilities or special commitments to American Indians, Alaska Natives, and affiliated island communities to help them prosper." DOI has a wide range of responsibilities including, among other things, mapping, geological, hydrological, and biological science; migratory bird, wildlife, and endangered species conservation; surface-mined lands protection and restoration; and historic preservation.[89]

Because final FY2019 funding was not available at the time the FY2020 budget was prepared, requested R&D funding is compared to the FY2018 actual funding.

The Administration is requesting $12.6 billion in net discretionary funding for DOI in FY2020.[90] Of that amount, $757 million is requested for R&D funding, $148 million (16.3%) below the FY2018 actual level of

---

[87] NOAA, *Budget Estimates Fiscal Year 2020*, 2019.

[88] This section was written by Laurie Harris, Analyst in Science and Technology Policy, CRS Resources, Science, and Industry Division.

[89] Department of the Interior, *Strategic Plan for Fiscal Years 2018-2022* and *Strategic Plan for Fiscal Years 2014- 2018*, available at https://www.doi.gov/performance/strategic-planning.

[90] Department of the Interior, *Fiscal Year 2020: The Interior Budget in Brief*, March 2019, p. DH-6.

$905 million.[91] Of the President's FY2020 DOI R&D funding request, 8.9% is for basic research, 73.3% is for applied research, and 17.8% is for development. The U.S. Geological Survey (USGS) is the only DOI component that conducts basic research.[92]

Funding for DOI R&D is generally included in appropriations line items that also include non-R&D activities. How much of the funding provided in appropriations legislation is allocated to R&D specifically is unclear unless funding is provided at the precise level of the request. In general, R&D funding levels are known only after DOI components allocate their appropriations to specific activities and report those figures.

U.S. Geological Survey

The USGS accounts for approximately two-thirds of all DOI R&D funding. A single appropriations account, Surveys, Investigations, and Research (SIR), provides all USGS funding. USGS R&D is conducted under seven SIR activity/program areas: Ecosystems; Land Resources; Energy, Minerals, and Environmental Health; Natural Hazards; Water Resources; Core Science Systems; and Science Support.

The President's total FY2020 budget request for USGS is $984 million. Of this amount, $481 million would be for R&D, a decrease of $119 million (19.8%) from the FY2018 enacted level of $600 million.[93]

Other DOI Components

The President's FY2020 request also includes R&D funding for the following DOI components:[94]

[91] Email correspondence between the DOI and CRS, April 11, 2019.
[92] Ibid.
[93] Ibid.
[94] Ibid.

- Bureau of Reclamation (BOR): $84.0 million in applied research and development funding for FY2020, down $26.4 million (23.9%) from FY2018.
- Bureau of Ocean Energy Management (BOEM): $100.4 million in applied research and development funding for FY2020, up $22.1 million (28.2%) from FY2018—the only component that would receive an increase in R&D funding.
- Fish and Wildlife Service (FWS): $15.5 million in applied research for FY2020, down $17.2 million (52.5%) from FY2018.
- National Park Service (NPS): $25.9 million in applied research and development for FY2020, down $1.1 million (4.2%) from FY2018.

Table 16. Department of the Interior R&D (budget authority, in millions of dollars)

	FY2018 Actual	FY2020 Request	FY2020 House	FY2020 Senate	FY2020 Enacted
U.S. Geological Survey (USGS)	600.5	481.4			
Bureau of Reclamation (BOR)	110.5	84.0			
Bureau of Ocean Energy Management (BOEM)	78.3	100.4			
Fish and Wildlife Service (FWS)	32.7	15.5			
National Park Service (NPS)	27.0	25.9			
Bureau of Safety and Environmental Enforcement (BSEE)	26.7	24.5			
Bureau of Land Management (BLM)	20.9	19.0			
Bureau of Indian Affairs (BIA)	5.0	5.0			
Wildland Fire Management (WFM)	3.0	0.0			
Office of Surface Mining Reclamation and Enforcement (OSMRE)	0.5	1.5			
Department of the Interior, R&D Total	$905.0	$757.2			

Source: Email correspondence between the DOI and CRS, April 11, 2019.

Notes: Totals may differ from the sum of the components due to rounding. Figures for the columns headed "FY2020 House," "FY2020 Senate," and "FY2020 Enacted" will be added, if available, as each action is completed.

- Bureau of Safety and Environmental Enforcement (BSEE): $24.5 million in applied research for FY2020, down $2.2 million (8.2%) from FY2018.
- Bureau of Land Management (BLM): $19.0 million in applied research and development for FY2020, down $1.9 million (9.0%) from FY2018.
- Bureau of Indian Affairs (BIA): $5.0 million in applied research for FY2020, equal to the actual amount from FY2018.
- Wildland Fire Management (WFM): No funding requested for R&D for FY2020, down $3.0 million (100.0%) from FY2018.[95]
- Office of Surface Mining Reclamation and Enforcement (OSMRE): $1.5 million in applied research for FY2020, up $970,000 (190%) from FY2018.

Table 16 summarizes FY2018 actual R&D funding and the President's FY2020 R&D funding request for DOI components.

DEPARTMENT OF VETERANS AFFAIRS[96]

The Department of Veterans Affairs (VA) operates and maintains a national health care delivery system to provide eligible veterans with medical care, benefits, and social support. As part of the agency's mission, it seeks to advance medical R&D in areas most relevant to the diseases and conditions that affect the health care needs of veterans.

The President is proposing $1.4 billion for VA R&D in FY2020, an increase of $12 million (1%) from FY2019. (See Table 17). VA R&D is funded through two accounts—the Medical and Prosthetic Research account and the Medical Care Support account. The Medical Care Support account also includes non-R&D funding, and the amount of funding that

[95] The FY2020 budget request for the Wildland Fire Management Program is $919.9 million (non-R&D funding).

[96] This section was written by Marcy E. Gallo, Analyst in Science and Technology Policy, CRS Resources, Science, and Industry Division.

will be allocated to support R&D through appropriations legislation is unclear unless funding is provided at the precise level of the request. In general, R&D funding levels from the Medical Care Support account are only known after the VA allocates its appropriations to specific activities and reports those figures. The Medical Care Support account provides administrative and other support for VA researchers and R&D projects, including infrastructure maintenance. The FY2020 request includes $762 million for VA's Medical and Prosthetic Research account, a decrease of $17 million (2%), and $648 million in funding for research supported by the agency's Medical Care Support account, an increase of $29 million (5%).

According to the President's request, FY2020 strategic priorities for VA R&D include increasing the access of veterans to clinical trials; increasing the transfer and translation of VA R&D; and "transforming VA data into a national resource" by reducing the time and effort needed to appropriately access, properly understand, and effectively use VA data for research. Clinical priorities for VA R&D in FY2020 include efforts to treat veterans at risk of suicide and research to address chronic pain and opioid addiction, posttraumatic stress disorder, traumatic brain injury, and Gulf War illness.

The Medical and Prosthetics R&D program is an intramural program managed by the Veteran Health Administration's Office of Research and Development (ORD) and conducted at VA Medical Centers and VA-approved sites nationwide. According to ORD, the mission of VA R&D is "to improve Veterans' health and well-being via basic, translational, clinical, health services, and rehabilitative research and to apply scientific knowledge to develop effective individualized care solutions for Veterans."[97] ORD consists of four main research services each headed by a director:

---

[97] Department of Veterans Affairs, website, "Office of Research & Development," https://www.research.va.gov/about/ default.cfm.

- Biomedical Laboratory R&D conducts preclinical research to understand life processes at the molecular, genomic, and physiological levels.
- Clinical Science R&D supports clinical trials and other human subjects research to determine the feasibility and effectiveness of new treatments such as drugs, therapies, or devices, compare existing therapies, and improve clinical care and practice.
- Health Services R&D conducts studies to identify and promote effective and efficient strategies to improve the quality and accessibility of the VA health system and patient outcomes, and to minimize health care costs.
- Rehabilitation R&D conducts research and develops novel approaches to improving the quality of life of impaired and disabled veterans.

Table 17. Department of Veterans Affairs R&D (budget authority, in millions of dollars)

Account	FY2019 Enacted	FY2020 Request	FY2020 House	FY2020 Senate	FY2020 Enacted
Medical and Prosthetic Research	$779.0	$762.0			
Medical Care Support	618.3	647.7			
Veterans Affairs, Total R&D	$1,397.3	$1,409.7			

Source: Department of Veterans Affairs, Volume II: Medical Programs and Information Technology Programs, Congressional Submission, FY2020 Funding and FY2021 Advance Appropriations, p. VHA-357, https://www.va.gov/budget/docs/summary/fy2020VAbudgetVolumeIImedicalProgramsAndInformationTechnology.pdf.

Notes: Totals may differ from the sum of the components due to rounding. Figures for the columns headed "FY2020 House," "FY2020 Senate," and "FY2020 Enacted" will be added, if available, as each action is completed. VA researchers also receive grants from other federal and nonfederal resources, including the National Institutes of Health, the Department of Defense, and the Centers for Disease Control and Prevention; these resources are estimated at $570 million in FY2019 and $570 million in FY2020. Additionally, the VA estimates reimbursements associated with agency R&D at $56 million in FY2019 and $55 million in FY2020, increasing the total amount of R&D performed at VA to $2.02 billion in FY2019 and $2.03 billion in the FY2020 request.

Table 18. Department of Veterans Affairs R&D by Designated Research Area (in millions of dollars)

Designated Research Area	FY2019	FY2020 Request
Acute and Traumatic Injury	$25.2	$25.5
Aging	140.0	141.5
Autoimmune, Allergic, and Hematopoietic Disorders	28.8	29.1
Cancer	57.3	57.9
Central Nervous System Injury and Associated Disorders	106.2	107.3
Degenerative Diseases of Bones and Joints	37.8	38.2
Dementia and Neuronal Degeneration	33.6	34.0
Diabetes and Major Complications	40.9	41.3
Digestive Diseases	17.6	17.8
Emerging Pathogens/Bio-Terrorism	1.6	1.6
Gulf War Veterans Illness	15.0	15.0
Health Systems	75.1	75.8
Heart Disease/Cardiovascular Health	71.5	72.2
Infectious Disease	31.3	31.6
Kidney Disorders	15.9	16.1
Lung Disorders	26.0	26.3
Mental Illness	119.1	120.4
Military Occupations and Environmental Exposures	22.9	23.1
Other Chronic Diseases	2.9	2.9
Prosthetics	19.1	19.3
Sensory Loss	15.1	15.3
Special Populations	26.8	27.1
Substance Abuse	33.0	33.3

Source: Department of Veterans Affairs, Volume II: Medical Programs and Information Technology Programs, Congressional Submission, FY2020 Funding and FY2021 Advance Appropriations, p. VHA-380, https://www.va.gov/budget/docs/summary/fy2020VAbudgetVolumeIImedicalProgramsAndInformationTechnology.pdf.

Notes: Projects that span multiple areas may be included in several Designated Research Areas (DRAs); therefore, the amounts depicted in this table total to more than the FY2019 amount and the FY2020 request for Medical and Prosthetic Research. Columns for "FY2020 House, "FY2020 Senate," and "FY2020 Enacted" are not included in this table as these figures will only be available after Congress completes the appropriations process and VA determines how much of the appropriated funds will be allocated to each DRA.

In addition to intramural support, VA researchers are eligible to obtain funding for their research from extramural sources, including other federal agencies, private foundations and health organizations, and commercial entities. According to the President's FY2020 budget request, these additional R&D resources are estimated at $570 million in FY2020. However, unlike other federal agencies, such as the National Institutes of Health and the Department of Defense, VA does not have the authority to support extramural R&D by providing research grants to colleges, universities, or other non-VA entities.

Table 17 summarizes R&D program funding for VA in the Medical and Prosthetic Research and the Medical Care Support accounts. Table 18 details amounts to be spent in Designated Research Areas (DRAs), which VA describes as "areas of importance to our veteran patient population." Funding for research projects that span multiple areas may be included in several DRAs; thus, the amounts in Table 18 total to more than the appropriation or request for VA R&D.

DEPARTMENT OF TRANSPORTATION[98]

The Department of Transportation (DOT) was established by the Department of Transportation Act (P.L. 89-670) on October 15, 1966. The primary purposes of DOT research and development activities as defined by Section 6019 of the Fixing America's Surface Transportation Act (P.L. 114-94) are improving mobility of people and goods; reducing congestion; promoting safety; improving the durability and extending the life of transportation infrastructure; preserving the environment; and preserving the existing transportation system.

Funding for DOT R&D is generally included in appropriations line items that also include non-R&D activities. The amount of the funding provided by appropriations legislation that is allocated to R&D is unclear unless funding is provided at the precise level of the request. In general, R&D funding levels are known only after DOT agencies allocate their final appropriations to specific activities and report those figures.

The Administration is requesting $1.089 billion for DOT R&D activities and facilities in FY2020, a decrease of $5.8 million (0.5%) from

[98] This section was written by Daniel Morgan, Specialist in Science and Technology Policy, CRS Resources, Science, and Industry Division.

FY2019. (See Table 19). Three DOT agencies—the Federal Aviation Administration (FAA), the Federal Highway Administration (FHWA), and the National Highway Traffic Safety Administration (NHTSA)—would account for over 90% of DOT R&D under the FY2020 request.

FEDERAL AVIATION ADMINISTRATION

The President's FY2020 request of $512.3 million for R&D activities and facilities at FAA would be an increase of $10.4 million (2.1%) from FY2019. The request includes $120 million for the agency's Research, Engineering, and Development (RE&D) account, a reduction of $71.1 million (37.2%) from FY2019. Funding within the RE&D account seeks to improve aircraft safety through research in fields such as fire safety, advanced materials, propulsion systems, aircraft icing, and continued airworthiness, in addition to safety research related to unmanned aircraft systems and the integration of commercial space operations into the national airspace.

Federal Highway Administration

According to the President's budget request

> FHWA's contributions to researching and implementing transformative innovations and technologies are changing the way roads, bridges, and other facilities are planned, designed, built, managed, and maintained across the country to be more responsive to current and future needs.[99]

---

[99] Department of Transportation, *U.S. Department of Transportation: Budget Highlights 2020*, p. 36, https://www.transportation.gov/sites/dot.gov/files/docs/mission/budget/333126/budgethighlightsfinal040519.pdf.

The President's request of $420 million for R&D activities and facilities at FHWA would be an increase of $39 million (10.2%) from FY2019. The request includes $125 million for FHWA's Highway Research and Development program, which seeks to improve safety, enhance the design and construction of transportation infrastructure, provide data and analysis for decision-making, and reduce congestion. The program supports highway research in such areas as the impact of automated driving systems, infrastructure durability, resilience, and environmental sustainability, and the factors that contribute to death and injury related to roadway design, construction, and maintenance. The request also includes $100 million for research to facilitate the development of a connected, integrated, and automated transportation system under the agency's Intelligent Transportation Systems program.

National Highway Traffic Safety Administration

The President is requesting $62.1 million in R&D and R&D facilities funding in FY2020 for NHTSA, $13.8 million (18.2%) below FY2019. NHTSA R&D focuses on automation and the study of human machine interfaces, advanced vehicle safety technology, ways of improving vehicle crashworthiness and crash avoidance, reducing unsafe driving behaviors, and alternative fuels vehicle safety.

OTHER DOT COMPONENTS

R&D activities are also supported by several other DOT components or agencies (see Table 19). The President's FY2020 request includes DOT R&D and R&D facilities funding for

- the Federal Railroad Administration (FRA), totaling $23.1 million, $21.6 million (48.3%) below the FY2019 level of $44.6 million;

- the Federal Transit Administration (FTA), totaling $28 million, $2 million (6.7%) below the FY2019 level of $30 million;
- the Pipeline and Hazardous Materials Safety Administration (PHMSA), totaling $21.5 million, $3 million (12.1%) below the FY2019 level of $24.5 million;
- the Office of the Secretary (OST), totaling $13.1 million, $14.8 million (53.2%) below the FY2019 level of $27.9 million; and
- the Federal Motor Carrier Safety Administration (FMCSA), totaling $9.1 million, the same amount as FY2019.

Table 19. Department of Transportation R&D Activities and Facilities (budget authority, in millions of dollars)

	FY2019 Enacted	FY2020 Request	FY2020 House	FY2020 Senate	FY2020 Enacted
Federal Aviation Administration	$501.9	$512.3			
Research, Engineering, and Development	191.1	120.0			
Federal Highway Administration	381.0	420.0			
Highway Research and Development	112.6	125.0			
Intelligent Transportation Systems	90.1	100.0			
National Highway Traffic Safety Administration	75.9	62.1			
Federal Railroad Administration	44.6	23.1			
Railroad Research and Development	40.6	19.0			
Federal Transit Administration	30.0	28.0			
Pipeline and Hazardous Materials Safety Administration	24.5	21.5			
Office of the Secretary	27.9	13.1			
Federal Motor Carrier Safety Administration	9.1	9.1			
DOT, R&D Total	$1,094.9	$1,089.1			

Sources: U.S. Department of Transportation, *Fiscal Year 2020 Budget Estimates*, https://www.transportation.gov/ mission/budget/fiscal-year-2020-budget-estimates.

Notes: Amounts include R&D and R&D facilities. Components may not add to total due to rounding. Lines in italics are components of the agency lines above them and are not counted separately in the total. Figures for the columns headed “FY2020 House,” “FY2020 Senate,” and “FY2020 Enacted” will be added, if available, as each action is completed.

DEPARTMENT OF HOMELAND SECURITY[100]

The Department of Homeland Security (DHS) has identified five core missions: to prevent terrorism and enhance security, to secure and manage the borders, to enforce and administer immigration laws, to safeguard and secure cyberspace, and to ensure resilience to disasters. New technology resulting from research and development can contribute to achieving all these goals. The Directorate of Science and Technology (S&T) has primary responsibility for establishing, administering, and coordinating DHS R&D activities. Other components, such as the Countering Weapons of Mass Destruction Office, the U.S. Coast Guard, and the Transportation Security Administration, conduct R&D relating to their specific missions.

The President's FY2020 budget request for DHS includes $438 million for activities identified as R&D. This would be a reduction of 31.6% from $640 million in FY2019. The total includes $303 million for the S&T Directorate and smaller amounts for six other DHS components. See Table 20.

The S&T Directorate performs R&D in several laboratories of its own and funds R&D performed by the DOE national laboratories, industry, universities, and others. It also conducts testing and other technology-related activities in support of acquisitions by other DHS components. The Administration's FY2020 request of $303 million for the S&T Directorate R&D account is a decrease of 40.7% from $511 million in FY2019. The request includes no funding for cybersecurity R&D ($89.1 million in FY2019), which would instead be conducted in the Cybersecurity Infrastructure Security Agency ($31 million for R&D in the FY2020 request, up from $13 million in FY2019). The remaining thrust areas in the S&T Directorate's Research, Development, and Innovation budget line would all decrease, by amounts ranging from 12.1% (Counter Terrorist) to 40.4% (Border Security). Funding for University Centers of Excellence would decrease from $37 million in FY2019 to $18 million in FY2020.

---

[100] This section was written by Daniel Morgan, Specialist in Science and Technology Policy, CRS Resources, Science, and Industry Division.

In addition to its R&D account, the S&T Directorate receives funding for laboratory facilities and other R&D-related expenses through its Operations and Support account (not shown in the Table). In this account, the FY2020 request for Laboratory Facilities is $116 million, down 4.9% from $122 million in FY2019. The Laboratory Facilities request includes no funding for the National Urban Security Technology Laboratory, which the Administration proposes to close, or for the National Bio and Agro-Defense Facility (NBAF), which the S&T Directorate is building using previously appropriated funds but will transfer to the USDA once it becomes operational. Requested funding in Laboratory Facilities for the National Biodefense Analysis and Countermeasures Center (NBACC) is $29 million, the same as in FY2019.

The request for R&D in the Countering Weapons of Mass Destruction Office is $68 million, down from $83 million in FY2019.

Table 20. Department of Homeland Security R&D Accounts (budget authority, in millions of dollars)

	FY2019 Enacted	FY2020 Request	FY2020 House	FY2020 Senate	FY2020 Enacted
Science and Technology Directorate	$511	$303			
Countering Weapons of Mass Destruction Office	83	68			
Transportation Security Administration	21	21			
U.S. Coast Guard	20	5			
Cybersecurity and Infrastructure Security Agency	13	31			
Office of the Under Secretary for Management	3	—			
U.S. Secret Service	3	11			
Total, DHS R&D	653	438			

Sources: FY2020 DHS congressional budget justification, https://www.dhs.gov/publication/congressional-budgetjustification-fy-2020.

Notes: Table includes accounts titled "Research and Development" in each DHS component. Some other accounts may also fund R&D-related activities. Some components may not add to totals due to rounding. Figures for the columns headed "FY2020 House," "FY2020 Senate," and "FY2020 Enacted" will be added, if available, as each action is completed.

ENVIRONMENTAL PROTECTION AGENCY[101]

The U.S. Environmental Protection Agency (EPA), the federal regulatory agency responsible for administering a number of environmental pollution control laws, funds a broad range of R&D activities to provide scientific tools and knowledge that support decisions relating to preventing, regulating, and abating environmental pollution. Since FY2006, Congress has funded EPA through the Interior, Environment, and Related Agencies appropriations acts.

Appropriations for EPA R&D are generally included in line-items that also include non-R&D activities. Annual appropriations bills and the accompanying committee reports do not identify precisely how much funding provided in appropriations bills is allocated to EPA R&D alone. EPA determines its R&D funding levels in operation through allocating its appropriations to specific activities and reporting those amounts.

The agency's Science and Technology (S&T) appropriations account[102] funds much of EPA's scientific research activities, which include R&D conducted by the agency at its own laboratories and facilities, and R&D and related scientific research conducted by universities, foundations, and other nonfederal entities that receive EPA grants. The S&T account receives a base appropriation and a transfer from the Hazardous Substance Superfund (Superfund) account for research on more effective methods remediating contaminated sites.

---

[101] This section was written by Robert Esworthy, Specialist in Environmental Policy, CRS Resources, Science, and Industry Division. For an overview of the President's FY2020 budget request and FY2019 enacted appropriations for EPA see CRS In Focus IF11153, *U.S. Environmental Protection Agency (EPA) Appropriations: FY2020 President's Budget Request*, by Robert Esworthy and David M. Bearden, and CRS In Focus IF11067, *U.S. Environmental Protection Agency (EPA) FY2019 Appropriations*, by Robert Esworthy and David M. Bearden.

[102] In 1995, Congress established eight statutory accounts for EPA, including the S&T account. The S&T account incorporates elements of the former EPA Research and Development account, as well as portions of the former Salaries and Expenses and Program Operations accounts, which were in place until FY1996. Currently, including the S&T account, discretionary funding is annually appropriated to EPA among 10 statutory accounts established by Congress over time in annual appropriations acts. Because of the differences in the scope of the activities included in these accounts, comparisons before and after FY1996 are not readily available.

EPA's Office of Research and Development (ORD) is the primary manager of R&D at EPA headquarters and laboratories around the country, as well as external R&D. A large portion of the S&T account funds EPA R&D activities managed by ORD, including research grants. Programs implemented by other offices within EPA also may have a research component, but the research component is not necessarily the primary focus of the program.

As with the President's FY2019 budget request, the FY2020 request proposes reductions and eliminations of funding for FY2020 across a number of EPA programs and activities. The President's FY2020 request includes a total of $6.07 billion for EPA,[103] $2.78 billion (31%) less than the total $8.85 billion FY2019 enacted appropriations for EPA (after rescissions) provided in Titles II and IV of Division E of the Consolidated Appropriations Act, FY2019 (P.L. 116-6), and $123.4 million (2%) less than the FY2019 request of $6.19 billion for EPA. The reductions proposed in the FY2020 request are distributed across EPA operational functions and activities as well as grants for states, tribes, and local governments. With the exception of the Building and Facilities account, the President's FY2020 request proposes funding reductions below FY2019 enacted levels for the nine other EPA appropriations accounts, although funding for some program areas within the accounts would remain constant or increase. Some Members of Congress expressed concerns regarding proposed reductions of funding for EPA scientific research programs during hearings on the President's FY2020 budget request.[104]

---

[103] U.S. EPA, *Fiscal Year 2020 Justification of Appropriations Estimates for the Committee on Appropriation, March 2019,* https://www.epa.gov/planandbudget/fy-2020-justification-appropriation-estimates-committee-appropriations.

[104] House Committee on Appropriations, Subcommittee on Interior, Environment, and Related Agencies, *Budget: Environmental Protection Agency*, hearing, April 2, 2019, https://appropriations.house.gov/legislation/hearings/budgetenvironmental-protection-agency; Senate Committee on Appropriations, Subcommittee on Interior, Environment, and Related Agencies, *Review of the FY2020 Budget Request for the Environmental Protection Agency*, hearing April 3, 2019, https://www.appropriations.senate.gov/hearings/review-of-the-fy2020-budget-request-for-the-environmentalprotection-agency; House Committee on Energy and Commerce, Subcommittee on Environment and Climate Change, *Hearing on "The Fiscal Year 2020 EPA Budget,"* hearing, April 9, 2019, https://energycommerce.house.gov/committee-activity/hearings/hearing-on-the-fiscal-year-2020-epa-budget.

Including a $17.8 million transfer from the Superfund account, the President's FY2020 budget request proposes $480.8 million for EPA's S&T account, [105] $241.1 million (33.4%) less than the FY2019 enacted $722.0 million which includes a $15.5 million transfer and $11.3 million account specific rescissions.[106] The FY2020 request would provide an increase (3.1%) compared to the FY2019 request of $466.4 million, which includes a $17.4 million transfer. The President's FY2020 request proposes a rescission for EPA but does not specify a rescission within the S&T or other appropriations accounts.[107] This accounting difference does not allow for direct comparisons of funding within EPA's S&T account including specific rescissions.

Table 21 at the end of this section includes the President's FY2020 request for program areas and activities within EPA's S&T account as presented in EPA's *FY2020 Congressional Budget Justification* compared to the FY2019 enacted appropriations as reported in the Conference Report (H.Rept. 116-9) accompanying the FY2019 consolidated appropriations that includes the Department of Interior, Environment, and Related Agencies appropriations.[108]

Consistent with other recent House and Senate Appropriations Committee reports and explanatory statements, the conference report H.Rept. 116-9 accompanying the FY2019 enacted appropriations did not specify funding for all subprogram areas reported in EPA's budget

---

[105] U.S. EPA, *Fiscal Year 2020 Justification of Appropriations Estimates for the Committee on Appropriations*, March 2019; see "Science and Technology" pp. 48-144, 812-813, and "Eliminated Programs" pp. 822-827, https://www.epa.gov/planandbudget/fy-2020-justification-appropriation-estimates-committee-appropriations.

[106] For FY2019, Title II of Div. E, P.L. 116-6 rescinded $210.5 million of prior-year funds specifying proportional allocations from EPA's S&T account (see end of **Table 21** as well as the Environmental Programs and Management (EPM) and the State and Tribal Assistance Grants (STAG) accounts).

[107] The President's FY2020 request proposes an agency-wide rescission of prior-year funds totaling $227.0 million for EPA but does not specify amounts within the individual appropriations accounts, including the S&T account. The reduction is referred to as agency-wide "cancellation of funds" in the President's FY2020 request and EPA's *Fiscal Year 2020 Justification of Appropriations Estimates for the Committee on Appropriation*.

[108] U.S. Congress, Conference Committee, *Making Further Continuing Appropriations for the Department of Homeland Security for Fiscal Year 2019, and for Other Purposes*, conference report to accompany H.J.Res. 31, 116th Cong., 1st Sess., H.Rept. 116-9 (Washington, DC: GPO, 2019).

justification. S&T subprogram areas not reported in congressional reports and statements are noted in the Table 21 as "NR" (not reported). Additionally, the President's FY2018, FY2019, and FY2020 requests and EPA's congressional budget justifications have modified the titles for some of the program areas relative to previous Administrations' budget requests and congressional committee reports' presentations. The House and Senate Appropriations Committees have generally adopted the modified program area titles as presented in the recent budget requests.

During House and Senate Committee hearings regarding the President's FY2020 budget request for EPA,[109] Members generally did not support a number of the proposed reductions and eliminations of funding for EPA, including proposed reductions in funding for scientific research programs. Reductions proposed in the FY2020 budget request below the FY2019 enacted levels were distributed across EPA operational functions and activities as well as grants for states, tribes, and local governments.

As shown in Table 21, with few exceptions the requested FY2020 amount for the S&T account for individual EPA program area and activity line items would be less than the FY2019 enacted appropriations. The FY2020 request did not propose to completely eliminate funding for the broader program areas; however, eliminations (no funding is requested for FY2020) are proposed for line-item activities below the program areas as indicated in Table 21. These program areas include

- Atmospheric Protection Program (formerly GHG [greenhouse gas] Reporting Program and Climate Protection Program),
- Indoor Air Radon Program, and ▫ Reduce Risks from Indoor Air.

For other program areas, proposed reductions in funding included eliminations of certain programs. For example, the proposed reduction in funding for Research: Air and Energy, Research: Safe and Sustainable Water Resources, Research: Sustainable and Healthy Communities, and

[109] See footnote 104.

Research: Chemical Safety and Sustainability program areas for FY2020[110] included the proposed elimination of funding for the Science to Achieve Results (STAR) program.[111]

P.L. 116-6 included $5.0 million for Research: National Priorities within the S&T account for FY2019,[112] an increase compared to $4.1 million included for FY2018. As in the previous Administration's fiscal year requests, the President's FY2020 budget request did not include funding for Research: National Priorities.[113]

In addition to clarifying certain funding allocations within the S&T account and consistent with the prior fiscal year appropriations committee reports and explanatory statements, H.Rept. 116-9 provided additional guidance for certain program areas and activities within the S&T account for FY2019. Topics expressly referenced included

- Alternative Testing;
- Computational Toxicology;
- Enhanced Aquifer Use;
- Integrated Risk Information System (IRIS);
- Nanomaterials Research;
- Innovative Research Partnerships;
- Intramural Animal Testing;
- Science to Achieve Results (STAR) Grants;

[110] For a description of the activities included under this program activity within EPA's S&T account see the agency's FY2020 Congressional budget justification, U.S. EPA *Fiscal Year 2020 Justification of Appropriations Estimates for the Committee on Appropriation*, pp. 115-120.

[111] See discussion under the heading "FY 2020 Change from FY 2019 Annualized Continuing Resolution (Dollars in Thousands)" within this program area in U.S. EPA *Fiscal Year 2020 Justification of Appropriations Estimates for the Committee on Appropriation*, p. 113, p.120, p. 126, and p. 132. See also, discussion under "Eliminated Programs" pp. 822-827.

[112] The grants were to be independent of the Science to Achieve Results (STAR) grant program. The grants are subject to a 25% matching funds requirement as stipulated in "Explanatory Statement" submitted by the Chairman of the House Committee on Appropriations in the *Congressional Record*, vol. 164, no. 50-Book II (March 22, 2018), p. H2621 and pp. H2665-H2666.

[113] Referred to as "Congressional Priorities" in U.S. EPA *Fiscal Year 2020 Justification of Appropriations Estimates for the Committee on Appropriation*. Not requesting funding for this program is consistent with the previous Administration's fiscal year budget requests.

Table 21. U.S. Environmental Protection Agency Science and Technology (S&T) Account (appropriations, in millions of dollars)

S&T Program Areas/Activities	FY2019 Enacted	FY2020 Request	FY2020 House	FY2020 Senate	FY2020 Enacted
Clean Air [and Climate][a]	$116.5	$87.3			
Clean Air Allowance Trading Program	NR	5.7			
Atmospheric Protection Program [GHG (greenhouse gas) Reporting Program; and Climate Protection Program][a]	8.0	0.0			
Federal Support for Air Quality Management	NR	3.8			
Federal Vehicle and Fuel Standards and Certification	NR	77.8			
Enforcement (Forensics Support)	13.7	10.9			
Homeland Security	33.1	32.8			
Critical Infrastructure Protection	NR	7.5			
Preparedness, Response, and Recovery	NR	24.8			
Protection of EPA Personnel and Infrastructure	NR	0.5			
Indoor Air and Radiation	6.0	4.8			
Indoor Air: Radon Program	NR	0.0			
Radiation: Protection	NR	1.0			
Radiation: Response Preparedness	NR	3.8			
Reduce Risks from Indoor Air	NR	0.0			
Information Technology/Data Management/Security	3.1	2.7			
Operations and Administration	68.3	73.3			
Facilities Infrastructure and Operations	NR	67.2			
Workforce Reshaping[b]	—	6.0			
Pesticide Licensing	6.0	5.3			
Protect Human Health from Pesticide Risk	NR	2.4			
Protect the Environment. from Pesticide Risk	NR	2.3			
Realize the Value of Pesticide Availability	NR	0.6			
Research: Air [Climate] and Energy[a]	94.9	31.7			
Research: Chemical Safety and Sustainability	126.9	86.6			
Human Health Risk Assessment	NR	22.7			
Research: Computational Toxicology	21.4	17.6			
Research: Endocrine Disruptor	16.3	10.3			
Research: Other Activities	NR	35.9			
Research: Safe and Sustainable Water Resources	106.3	70.0			
Research: Sustainable and Healthy Communities	134.3	53.6			
Water: Human Health Protection (Drinking Water Programs)	3.5	4.1			

S&T Program Areas/Activities	FY2019 Enacted	FY2020 Request	FY2020 House	FY2020 Senate	FY2020 Enacted
Research: National/Congressional Priorities (Water Quality and Support Grants)[c]	5.0	0.0			
Subtotal Base Appropriations	717.7	463.1			
Transfer in from Hazardous Substance Superfund Account	15.5	17.8			
Total Appropriations Prior to Rescissions	733.2	480.8			
S&T Account Specific Rescission[d]	(11.3)[d]	NR[e]			
Total (Net Appropriations)	722.0	480.8			

Source: Prepared by CRS using information from the Congressional Record; House, Senate committee reports and explanatory statements; and EPA's Fiscal Year 2020 Justification of Appropriations Estimates for the Committee on Appropriations, March 2019.

Notes: Totals may differ from the sum of the components due to rounding. Figures for the columns headed "FY2020 House," "FY2020 Senate," and "FY2020 Enacted" will be added, if available, as each action is completed. NR (not reported) indicates those instances where funding or rescission ("cancelled") amounts were not specified.

a. Brackets [ ] denote title language as presented in previous Administrations' EPA budget justifications and congressional reports/explanatory statements.

b. This program activity was included in multiple EPA accounts in the FY2018, FY2019, and FY2020 budget justifications and had not been included in previous fiscal year EPA budget justifications.

c. Referred to as "Congressional Priorities" in the FY2020 and previous Administrations' budget justifications.

d. The Conference Report (H.Rept. 116-9) accompanying the FY2019 appropriations noted that EPA's current workforce was below prior levels and therefore included separate rescissions within the S&T and the Environmental Programs and Management (EPM) accounts to "capture expected savings associated with such changes." As specified in the act, the rescission within the S&T account is not to be applied to "Research: National Priorities."

e. The President's FY2020 request included a $227.0 million rescission of unobligated balances for EPA appropriations overall but did not specify a proportional allocation of the rescission by EPA accounts.

- Harmful Algal Blooms;
- Water Distribution Systems; and
- Water Security Test Beds.[114]

---

[114] U.S. Congress, Conference Committee, *Making Further Continuing Appropriations for the Department of Homeland Security for Fiscal Year 2019, and for Other Purposes*, conference report to accompany H.J.Res. 31, 116th Cong., 1st sess., H.Rept. 116-9 (Washington, DC: GPO, 2019), pp. 736-737. H.Rept. 116-9 refers to H.Rept. 115-765 and S.Rept. 115-276 for specific guidance regarding the administration of appropriated funds. The two other reports also include "Additional Guidance."

The size and structure of the agency's workforce, as was the case during consideration for the FY2018 and FY2019 appropriations, has been a topic of debate during congressional committee hearings regarding EPA's FY2020 appropriations.[115] Workforce reshaping was introduced in the FY2018 request and described as agency-wide organizational restructuring, "reprioritization of agency activities," and reallocation of resources.[116] The FY2020 request for the Operations and Administration program area within the S&T account includes $6.0 million for agency workforce reshaping and efforts to improve the management of EPA's laboratories.[117] As with the FY2018 enacted appropriations, P.L. 116-6 did not fund the President's FY2019 request for EPA workforce reshaping for FY2019.

EPA's reported proposed reorganizing strategies, potentially impacting certain aspects of EPA's Office of Research Development (ORD) and the operations of the EPA Office of the Science Advisor (OSA), as well as current EPA laboratories including the National Exposure Research Laboratory (NERL), the National Health and Environmental Effects Research Laboratory (NHEERL), and the National Risk Management Research Laboratory (NRMRL), have also been of interest to some Members of Congress.[118]

APPENDIX A. ACRONYMS AND ABBREVIATIONS

ACF	Administration for Children and Families
AFRI	Agriculture and Food Research Initiative
AHRQ	Agency for Healthcare Research and Quality

[115] See footnote 104.

[116] U.S. EPA, *Fiscal Year 2018 Justification of Appropriations Estimates for the Committee on Appropriations*, May 2017, https://www.epa.gov/planandbudget/fy-2018-justification-appropriation-estimates-committee-appropriations.

[117] See U.S. EPA *Fiscal Year 2020 Justification of Appropriations Estimates for the Committee on Appropriation*, pp. 92-96.

[118] U.S. EPA, "EPA's Office of Research and Development Reorganizes to Better Support EPA's Mission," April 8, 2019, press release, https://www.epa.gov/sciencematters/epas-office-research-and-development-reorganizes-bettersupport-epas-mission.

AI	Artificial Intelligence
AIMS	Arctic Infrastructure Modernization for Science
AMP	Advanced Manufacturing Partnership – or – Accelerating Medicines Partnership
AOAM	Agency Operations and Award Management
APHIS	Animal and Plant Health Inspection Service
ARPA-E	Advanced Research Projects Agency-Energy
ARS	Agricultural Research Service
B&F	Buildings and Facilities
BA	Budget Authority
BIA	Bureau of Indian Affairs
BLM	Bureau of Land Management
BOEM	Bureau of Ocean Energy Management
BOR	Bureau of Reclamation
BRAIN	Brain Research through Advancing Innovative Neurotechnologies
BSEE	Bureau of Safety and Environmental Enforcement
CDC	Centers for Disease Control and Prevention
CLARREO	Climate Absolute Radiance and Refractivity Observatory
CMS	Centers for Medicare and Medicaid Services
CR	Continuing Resolution
CRF	Construction of Research Facilities
DHP	Defense Health Program
DHS	Department of Homeland Security
DOC	Department of Commerce
DOD	Department of Defense
DOE	Department of Energy
DOI	Department of the Interior
DOT	Department of Transportation
DRA	Designated Research Area
EHR	Education and Human Resources
EOP	Executive Office of the President

EPA	Environmental Protection Agency
EPM	Environmental Programs and Management
EPSCoR	Experimental Program to Stimulate Competitive Research –or– Established Program to Stimulate Competitive Research
ERS	Economic Research Service
FAA	Federal Aviation Administration
FDA	Food and Drug Administration
FHWA	Federal Highway Administration
FIC	Fogarty International Center
FMCSA	Federal Motor Carrier Safety Administration
FRA	Federal Railroad Administration
FTA	Federal Transit Administration
FWS	Fish and Wildlife Service
FY	Fiscal Year
GDP	Gross Domestic Product
GHG	greenhouse gas
GRF	Graduate Research Fellowship
GWOT	Global War on Terror
HBCU	Historically Black Colleges and Universities
HEAL	Helping to End Addiction Long-Term
HHS	Department of Health and Human Services
HRSA	Health Resources and Services Administration
ICs	Institutes and Centers
IDeA	Institutional Development Award
IFF	Iraqi Freedom Fund
IRIS	Integrated Risk Information System
ISS	International Space Station
ITER	International Thermonuclear Experimental Reactor
ITS	Industrial Technology Services
JWST	James Webb Space Telescope
LEO	Low Earth Orbit
LHHS	Labor, HHS, and Education appropriations act

LSST	Large Synoptic Survey Telescope
MEP	Manufacturing Extension Partnership
MGI	Materials Genome Initiative
MREFC	Major Research Equipment and Facilities Construction
NASA	National Aeronautics and Space Administration
NASS	National Agricultural Statistics Service
NBACC	National Biodefense Analysis and Countermeasures Center
NBAF	National Bio and Agro-Defense Facility
NCATS	National Center for Advancing Translational Sciences
NCCIH	National Center for Complementary and Integrative Health
NCI	National Cancer Institute
NEI	National Eye Institute
NERL	National Exposure Research Laboratory
NESDIS	National Environmental Satellite, Data, and Information Service
NHEERL	National Health and Environmental Effects Research Laboratory
NHGRI	National Human Genome Research Institute
NHLBI	National Heart, Lung, and Blood Institute
NHTSA	National Highway Traffic Safety Administration
NIA	National Institute on Aging
NIAAA	National Institute on Alcohol Abuse and Alcoholism
NIAID	National Institute of Allergy and Infectious Diseases
NIAMS	National Institute of Arthritis and Musculoskeletal and Skin Diseases
NIBIB	National Institute of Biomedical Imaging and Bioengineering

NICHD	National Institute of Child Health and Human Development
NIDA	National Institute on Drug Abuse
NIDCD	National Institute on Deafness and Other Communication Disorders
NIDCR	National Institute of Dental and Craniofacial Research
NIDDK	National Institute of Diabetes and Digestive and Kidney Diseases
NIEHS	National Institute of Environmental Health Sciences
NIFA	National Institute of Food and Agriculture
NIGMS	National Institute of General Medical Sciences
NIH	National Institutes of Health
NIIMBL	National Institute for Innovation in Manufacturing Biopharmaceuticals
NIMH	National Institute of Mental Health
NIMHD	National Institute on Minority Health and Health Disparities
NINDS	National Institute of Neurological Disorders and Stroke
NINR	National Institute of Nursing Research
NIRSQ	National Institute for Research on Safety and Quality
NIST	National Institute of Standards and Technology
NITRD	Networking and Information Technology Research and Development
NLM	National Library of Medicine
NMFS	National Marine Fisheries Service
NNI	National Nanotechnology Initiative
NNMI	National Network for Manufacturing Innovation
NOAA	National Oceanic and Atmospheric Administration
NOS	National Ocean Service

NPS	National Park Service
NRI	National Robotics Initiative
NRMRL	National Risk Management Research Laboratory
NRT	NSF Research Traineeships
NSB	National Science Board
NSET	Nanoscale Science, Engineering, and Technology (NSTC Subcommittee)
NSF	National Science Foundation
NSTC	National Science and Technology Council
NWS	National Weather Service
OAR	Oceanic and Atmospheric Research
OCO	Overseas Contingency Operations
OD	NIH Office of the Director
OIG	Office of the Inspector General
OMAO	Office of Marine and Aviation Operations
OMB	Office of Management and Budget
ORD	Office of Research and Development
OSMRE	Office of Surface Mining Reclamation and Enforcement
OST	Office of the Secretary of Transportation
OSTP	Office of Science and Technology Policy
PACE	Pre-Aerosol, Clouds, and Ocean Ecosystem
PCORTF	Patient-Centered Outcomes Research Trust Fund
PE	Program Element
PHMSA	Pipeline and Hazardous Materials Safety Administration
PHS	Public Health Service
PMI	Precision Medicine Initiative
R&D	Research and Development
RDT&E	Research, Development, Test, and Evaluation
RE&D	Research, Engineering, and Development
REE	Research, Education, and Economics

RPG	Research Project Grant
RRA	Research and Related Activities
S&T	Science and Technology
SIR	Surveys, Investigations, and Research
SLS	Space Launch System
STAG	State and Tribal Assistance Grants
STAR	Science to Achieve Results
STEM	Science, Technology, Engineering, and Mathematics
STEP	Supercritical Transformational Electric Power
STRS	Scientific and Technical Research and Services
TOA	Total Obligational Authority
USDA	Department of Agriculture
USGCRP	U.S. Global Change Research Program
USGS	U.S. Geological Survey
VA	Department of Veterans Affairs
WFIRST	Wide Field Infrared Space Telescope
WFM	Wildland Fire Management

In: Key Congressional Reports for May 2019 ISBN: 978-1-53616-335-3
Editor: Aston Dunkley

Chapter 4

U.S. AIRBORNE ELECTRONIC ATTACK PROGRAMS: BACKGROUND AND ISSUES FOR CONGRESS*

John R. Hoehn

ABSTRACT

U.S. airborne electronic warfare (EW) programs involve developing and procuring EW aircraft and EW systems that are mounted on U.S. aircraft. The President's FY2020 budget request for the Department of Defense (DOD) proposes funding for a number of airborne EW programs.

The Role of Airborne EW in Modern Warfare

EW is a component of modern warfare, particularly in response to threats posed by potential adversaries such as Russia or China. EW refers to operations that use the electromagnetic spectrum (i.e., the "airwaves") to detect, listen to, jam, and deceive (or "spoof") enemy radars, radio communication systems, data links, and other electronic systems. EW

* This is an edited, reformatted and augmented version of Congressional Research Service, Publication No. R44572, dated May 14, 2019.

also refers to operations that defend against enemy attempts to do the same.

The shift in the international security environment from the post-Cold War era to an era of renewed great power competition has led to an increased focus on EW in U.S. defense planning and programming, particularly aspects of EW related to high-end warfare.

U.S. Airborne Electronic Attack Capabilities

Airborne EW capabilities are a component of U.S. military airpower. Although dedicated U.S. EW aircraft are relatively few in number compared with U.S. fighters, strike fighters, and attack aircraft, they play a role in helping to ensure the combat survivability and effectiveness of other aircraft and friendly forces on the ground.

DOD's three primary manned EW electronic attack aircraft are the Navy EA-18G Growler, the Air Force EC-130H Compass Call, and the Air Force EC-37B Compass Call Re-Host. A fourth manned aircraft—the F-35 Joint Strike Fighter—has extensive, integrated EW capabilities. DOD's primary airborne electronic attack payloads include the AN/ALQ-99 electronic attack suite, the Next Generation Jammer, and the Miniature Air Launched Decoy-Jammer.

EW Oversight Issues for Congress

Congress has continually shown interest in EW, and the decisions it makes regarding EW could affect future U.S. military capabilities and funding requirements. In particular, EW programs pose several potential oversight issues for Congress

- Whether DOD is prioritizing appropriately airborne EW programs in its planning and budgeting relative to other U.S. military EW programs (such as those for U.S. ground forces or Navy surface ships) and to other DOD non-EW priorities.
- Whether DOD's proposed mix of airborne EW capabilities and investments is appropriate.
- The evolution of technology and how new technologies can be employed for EW operations.
- The Air Force's planned rate for procuring EC-37Bs and replacing EC-130Hs.

INTRODUCTION[1]

This chapter focuses on selected U.S. airborne electronic attack programs. Such programs involve developing and procuring both the aircraft whose primary mission is electronic warfare (EW) and the EW systems that are mounted on U.S. aircraft. The President's FY2020 budget request for the Department of Defense (DOD) seeks funding for a number of airborne EW programs.

These programs pose a number of potential oversight issues for Congress, and its decisions on these issues could affect future U.S. military capabilities and funding requirements.

Congress has continually shown interest in EW, and airborne electronic attack in particular. Some Members have formed the EW Working Group, and they routinely discuss improving EW capabilities. The National Defense Authorization Acts over the past several years have included provisions related to EW and electronic attack. Most recently the FY2019 John S. McCain National Defense Authorization Act, discussed the Air Force's acquisition strategy for a new EW attack aircraft as well as a study to catalogue all EW capabilities.

BACKGROUND

Electronic Warfare

Overview[2]

Electronic warfare (EW)—sometimes also called electromagnetic maneuver warfare (EMW)[3]—is a component of modern warfare, particularly in response to threats posed by technologically sophisticated

[1] This chapter was originally authored by Ray Zuniga, then-Air Force Fellow.

[2] For a brief overview of EW, see CRS In Focus IF11118, *Defense Primer: Electronic Warfare*, by John R. Hoehn.

[3] See, for example, John Joyce, "Navy Expands Electromagnetic Maneuver Warfare for 'Victory at Sea,'" *Navy News Service*, November 2, 2017; Robert K. Ackerman, "Electromagnetic Maneuver Warfare Looms as New U.S. Navy Discipline," *Signal*, February 11, 2015.

potential adversaries such as Russia and China. EW generally refers to operations that use the electromagnetic spectrum (i.e., the "airwaves") to detect, listen to, jam, and deceive (or "spoof") enemy radars, radio communication systems and data links, and other electronic systems. It also refers to operations for defending against enemy attempts to do the same. More formally, DOD defines electronic warfare as "military action involving the use of electromagnetic and directed energy to control the electromagnetic spectrum or to attack the enemy."[4]

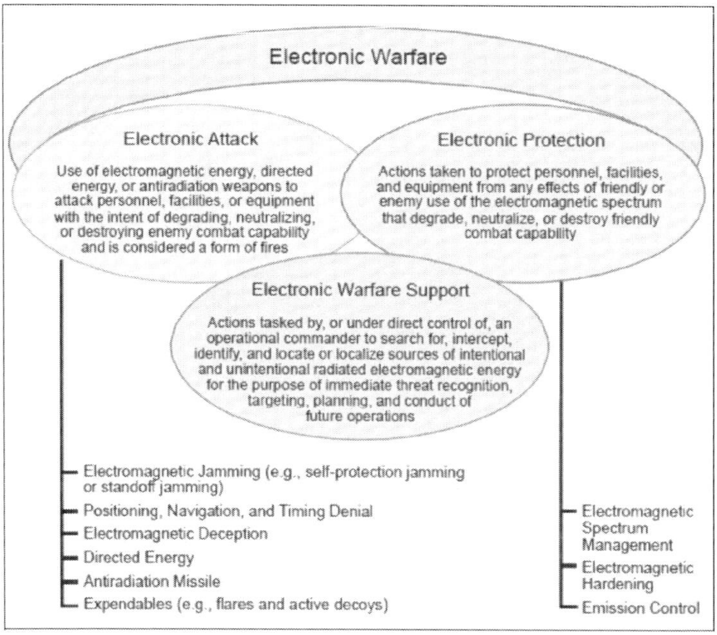

Source: Department of Defense, *Joint Publication 3-13.1, Electronic Warfare*, February 8, 2012, Figure I-3 on p. I-5, accessed March 13, 2019, at https://www.globalsecurity.org/military/library/policy/dod/joint/jp3_13_1_2012.pdf and https://publicintelligence.net/jcs-ew/.

Figure 1. Overview of Electronic Warfare.

---

[4] Department of Defense, *DOD Dictionary of Military and Associated Terms As of February 2019*, p. 78, accessed March 13, 2019, at https://www.jcs.mil/Portals/36/Documents/Doctrine/pubs/dictionary.pdf. See also Department of Defense, *Joint Publication 3-13.1, Electronic Warfare*, February 8, 2012, 144 pp. (including covers), accessed March 13, 2019, at https://www.globalsecurity.org/military/library/policy/dod/joint/jp3_13_1_2012.pdf and https://publicintelligence.net/jcs-ew/.

As shown in Figure 1, DOD divides EW into electronic warfare support, electronic protection, and electronic attack. *Electronic warfare support*, sometimes also referred to as electronic support measures (ESM), involves listening to an adversary's radar and radio transmissions in order to detect, locate, and understand how to avoid, jam, or deceive those systems. *Electronic protection* involves limiting the electromagnetic signatures of one's own military equipment and hardening one's own military equipment against the effects of enemy EW operations. *Electronic attack* (EA) involves jamming and deceiving enemy radars and radio communications and data links.

Developing ever-better EW systems is a component of the overall competition in military capabilities between major military powers. This issue is not frequently discussed publicly in much detail, because the specifics of EW programs tend to be classified and are closely related to intelligence systems and capabilities.

EW in an Era of Renewed Great Power Competition

During the Cold War, EW capabilities supported the overall competition in military capabilities between the U.S.-led NATO alliance and the Soviet-led Warsaw Pact alliance. The end of the Cold War and the shift in the early 1990s to the post-Cold War era—a period that featured reduced tensions between major powers and a strong U.S. military emphasis on countering terrorist and insurgent organizations—may have led to a reduced emphasis in U.S. defense plans and programs related to so-called high-end warfare, meaning high-intensity warfare against technologically sophisticated adversaries.

In recent years, the shift in the international security environment from the post-Cold War era to an era of renewed great power competition has increased the focus on EW in U.S. defense planning and programming.[5] In particular, attention has been given to aspects of EW related to high-end warfare and to concerns among some observers that the United States

[5] For more on this shift, see CRS Report R43838, *A Shift in the International Security Environment: Potential Implications for Defense—Issues for Congress*, by Ronald O'Rourke, and CRS In Focus IF11139, *Evaluating DOD Strategy: Key Findings of the National Defense Strategy Commission*, by Kathleen J. McInnis.

needs to strengthen its efforts in EW as part of its overall effort to preserve U.S. qualitative military superiority over potential adversaries such as Russia and China.

DOD notes Russia has placed an emphasis on EW in its military modernization effort.[6] For example, Russia reportedly has employed EW as part of its military operations in Ukraine and Syria.[7] DOD similarly states that China recognizes the importance of EW in modern military operations and is developing its EW capabilities as an integral part of its broad-based military modernization effort.[8] As China encourages greater integration between its civil and military technological and industrial bases,[9] its EW capabilities may benefit from the sophistication of its extensive civilian electronics industry.[10]

Relationship of EW to Cyberwarfare

EW emerged in the early and middle decades of the 20th century with the invention and spread of radio and radar and their use in military operations. It therefore predates cyberwarfare, which emerged decades later with the invention and spread of computers and the internet. Today, some overlap exists between EW and cyberwarfare, though there is a key

---

[6] See Department of Defense, Defense Intelligence Agency, *Russia Military Power, Building a Military to Support Great Power Aspirations*, 2017, pp. 32, 42.

[7] See, for example, Yuri Lapaiev, "Ukraine as Clandestine Testing Ground for Russian Electronic Warfare," *Eurasia Daily Monitor*, vol. 15, issue 157, November 5, 2018; "Russia Deploys Electronic Warfare in Syria," *Army Recognition*, October 17, 2018; Tom O'Connor, "Russia's Electronic Warfare System in Syria Will Be Able to Track Planes in Europe and Israel," *Newsweek*, September 28, 2018 Lara Seligman, "Russian Jamming Poses a Growing Threat to U.S. Troops in Syria," *Foreign Policy*, July 30, 2018 Roger N. McDermott, *Russia's Electronic Warfare Capabilities to 2025, Challenging NATO in the Electromagnetic Spectrum*, International Centre for Defence and Security, Tallinn, Estonia, September 2017, p. 39.

[8] See Department of Defense, *Annual Report to Congress: Military and Security Developments Involving the People's Republic of China 2018*, p. 74. For more on China's military modernization effort, see CRS Report R44196, *The Chinese Military: Overview and Issues for Congress*, by Ian E. Rinehart, and CRS Report RL33153, *China Naval Modernization: Implications for U.S. Navy Capabilities—Background and Issues for Congress*, by Ronald O'Rourke.

[9] For more on these efforts, see Department of Defense, *Annual Report to Congress: Military and Security Developments Involving the People's Republic of China 2018*, pp. 1-2, 81, 84-85, 121.

[10] For an example of these types of activities, see CRS In Focus IF10119, *U.S.-China Relations*, by Susan V. Lawrence, Wayne M. Morrison, and Jonah Langan-Marmur.

difference between the two. EW focuses on military operations that use the electromagnetic spectrum against radars and radio communication and data links, while cyberwarfare activities—which occur on a day-to-day basis, as well as during overt conflicts—target computers and servers, and involve significant use of the wired connections between them. EW and cyberwar activities can support one another.[11]

EW as an Element of U.S. Airpower

Although dedicated U.S. EW aircraft are relatively few in number compared with the number of U.S. fighters, strike fighters, and attack aircraft, they play a role in helping to ensure the combat survivability and effectiveness of other aircraft and friendly forces on the ground. EW aircraft detect and jam enemy radars and air defense command-and-control equipment, so that U.S. fighters, strike fighters, attack aircraft, and bombers can more safely penetrate enemy airspace. EA-18G Growlers (discussed below) accompany U.S. fighters, strike fighters, and attack aircraft on missions to penetrate enemy airspace. Other U.S. EW aircraft, such as the EC-130H Compass Call aircraft (discussed below), perform their EW missions from standoff locations in less contested airspace.

Fifth-generation stealthy U.S. aircraft such as the F-22 Raptor and the F-35 Joint Strike Fighter are less dependent on EW support than are less stealthy, earlier-generation U.S. aircraft. Even F-22s and F-35s, however, still benefit from EW support under certain circumstances.[12] EW aircraft support the Navy's Naval Integrated Fires Counter-Air (NIF-CA) concept

---

[11] DOD states that [s]ince cyberspace requires both wired and wireless links to transport information, both offensive and defensive cyberspace operations may require use of the EMS [electromagnetic spectrum] for the enabling of effects in cyberspace. Due to the complementary nature and potential synergistic effects of EW and CNO [computer network operations], they must be coordinated to ensure they are applied to maximize effectiveness. Cyberspace operations may be used to force an adversaryfrom wired to wireless networks that are vulnerable to EA [electronic attack]. EW may be used to set favorable conditions for cyberspace operations by stimulating networked sensors, denying wireless networks, or other related actions. In the defensive environment, EW systems may detect and defeat attacks across wireless access points. (Department of Defense, *Joint Publication 3-13.1, Electronic Warfare*, February 8, 2012, p. I-15, accessed March 13, 2019, at https://www.globalsecurity.org/military/library/policy/dod/joint/ jp3_13_1_2012.pdf and https://publicintelligence.net/jcs-ew/).

[12] The F-35 includes inherent EW capabilities, as discussed below.

and help ensure the combat survivability and effectiveness of less stealthy, earlier-generation U.S. aircraft and friendly forces on the ground.

EW Aircraft

Although various U.S. manned and unmanned aircraft perform EW operations, this chapter focuses on DOD's three primary manned EW electronic attack aircraft: the EA-18G Growler, the EC-130H Compass Call, and the EC-37B Compass Call Re-Host. It also focuses on a fourth manned aircraft, the F-35 Joint Strike Fighter, which has extensive built-in EW capabilities. Each of these four aircraft is discussed briefly below.

Boeing EA-18G Growler

The Boeing EA-18G Growler (Figure 2) is a Navy carrier-capable EW aircraft. Its primary mission is to detect and jam enemy radars. Among the 60 or more aircraft in an aircraft carrier's embarked air wing, typically four or five are EA-18Gs. These aircraft are also operated by the Royal Australian Air Force (RAAF).

The EA-18G is the successor to the carrier-capable EA-6B Prowler, which was operated by both the Navy and Marine Corps. The EA-18G achieved initial operational capability (IOC) in September 2009,[13] and EA-18Gs have gradually replaced EA-6Bs. The final operational EA-6Bs, operated by the Marine Corps, were retired in March 2019.[14]

Unlike the EA-6B, which was a four-seat aircraft, the EA-18G has a crew of two. The EA-6B was an EW variant of the Navy and Marine Corps carrier-capable A-6 Intruder attack plane; similarly, the EA-18G is an EW variant of the Navy and Marine Corps carrier-capable F/A-18F Super Hornet strike fighter.

---

[13] Department of the Navy, *U.S. Navy Program Guide 2017*, p. 11.

[14] See, for example, Shawn Snow, "EA-6B Prowler, One of the Saltiest Warfighters in the Marine Corps, Retires," *Marine Corps Times*, March 8, 2019; Justin Katz, "Marine Corps to Complete Prowler Retirement in March," Inside Defense, November 1, 2018.

Source: Photo accompanying Megan Eckstein, "Electronic Warfare Executive Committee to Focus on Strategy, Acquisition," *USNI News*, March 18, 2015.

Figure 2. EA-18G Growler.

The EA-18G is equipped with an airborne electronic attack (AEA) avionics suite that has evolved from the EA-6B's Improved Capability III (ICAP III) AEA system. As discussed below, the EA-18G carries AN/ALQ-99 jamming pods, which are to be replaced by Next Generation Jammer jamming pods. The Navy states that "the EA-18G's electronic attack upgrades meet or exceed EA-6B Airborne... Electronic Attack capability to detect, identify, locate and suppress hostile [electromagnetic] emitters; provide enhanced connectivity to National, Theater and strike assets; and provide organic precision emitter targeting for employment of onboard suppression weapons to fulfill operational requirements."[15] The Navy further states that

> [t]he EA-18G provides full-spectrum airborne electronic attack (AEA) capabilities to counter enemy air defenses and communication networks, most notably anti-radiation missiles. These capabilities continue to be in high demand in overseas contingency operations, where

---

[15] *Department of Defense, Fiscal Year (FY) 2018 Budget Estimates, Navy, Justification Book Volume 1 of 4, Aircraft Procurement, Navy, Budget Activity 01−04*, May 2017, p. 1 (PDF page 69 of 242).

Growler operations protect coalition forces and disrupt critical command and control links.[16]

The Air Force does not operate an aircraft directly analogous to the EA-18G. The last such Air Force aircraft was the EF-111 Raven, an EW variant of the F-111 fighter. The Air Force retired the last of its EF-111s in 1998.

The Navy states that "the [EA-18G] inventory objective of 160 aircraft will support ten carrier-based squadrons, five active expeditionary squadrons, and one reserve squadron."[17] A total of 163 EA-18Gs were procured through FY2016, including a final procurement of 10 in FY2016. The Department of the Navy does not plan further procurement of EA-18Gs. EA-18Gs, like F/A18E/Fs, currently are receiving funding for a service life extension;[18] the Growler is expected to be replaced starting in the 2030s.[19]

Lockheed Martin EC-130H Compass Call[20]

The EC-130H Compass Call (Figure 3) is an EW aircraft based on a modified version of the C130 Hercules cargo aircraft. The EW system on the aircraft is called the Compass Call system. The Air Force states that the EC-130H "disrupts enemy command and control communications and limits adversary coordination essential for enemy force management." The Compass Call system employs offensive counter-information and electronic attack (or EA) capabilities in support of U.S. and Coalition tactical air, surface, and special operations forces.

---

[16] Department of the Navy, *U.S. Navy Program Guide 2017*, p. 10.

[17] Department of the Navy, *U.S. Navy Program Guide 2017*, p. 11.

[18] Research and development efforts to improve the capabilities of EA-18Gs in coming years are funded in Program Element (PE) 0604269N, EA-18 Squadrons—a line item in the Navy's research and development account. In the Navy's proposed FY2020 budget, this PE is line 112 in the Navy's research and development account.

[19] Department of Defense *Annual Aviation Inventory and Funding Plan: Fiscal Years (FY) 2019-2048*, p. 10. Accessed at: https://apps.dtic.mil/dtic/tr/fulltext/u2/1062648.pdf.

[20] Unless otherwise indicated, information in this section is taken from U.S. Air Force fact sheet, "EC-130H Compass Call," September 23, 2015, accessed March 15, 2019. Although this fact sheet is dated September 23, 2019, the Air Force states that its online fact sheets contain up-to-date information and statistics.

Source: U.S. Air Force official website, EC-130H Compass Call Fact Sheet, September 23, 2015, accessed April 1, 2019, at http://www.af.mil/AboutUs/FactSheets/Display/tabid/224/Article/104550/ec-130h-compass-call.aspx.

Figure 3. EC-130H Compass Call.

The EC-130H is operated by a crew of 14, most of whom are assigned to operate the aircraft's EW systems. The EC-130H can be considered a so-called "low-density, high-demand asset," meaning a specialized asset that exists in DOD in relatively low numbers but that DOD uses extensively. A February 2018 press report states that

> [t]he small, 14-aircraft EC-130H fleet has been flying since 1981—and near-constantly in the Afghanistan, Iraq, and Syrian conflicts, because of the unique capability it offers in communications jamming and electronic attack. It has been a key element in the fight against ISIS, an adversary that has adapted high technology to its tactics and strategy....
>
> EC-130Hs there have been deployed nonstop since 2002, the longest continuously deployed Air Force unit in the Afghanistan war.[21]

The EC-130H achieved IOC in 1983. EC-130Hs are being replaced over time by new EC-37B Compass Call Re-Host aircraft (see next section). The Air Force projects in its FY2020 budget submission that it will have 13 EC-130Hs and no EC-37Bs in service at the end of FY2019,

[21] Brian W. Everstine, "Electronic Warfare Isn't Going Away—But EC-130Hs Will," *Air Force Magazine*, February 2018.

and 12 EC-130Hs and one EC-37B in service at the end of FY2020. While EC-130Hs remain in service, the Air Force plans to modernize them to improve their capabilities and reduce their maintenance costs, which have been rising as the aircraft have aged.[22]

EC-37B Compass Call Re-Host Aircraft

Air Force plans call for replacing the service's EC-130Hs over time with a total of 10 new EC-37B Compass Call Re-Host aircraft (Figure 4).

The first EC-37B was procured in FY2018, two more were procured in FY2019, and the Air Force's proposed FY2020 budget requests $114.1 million for the procurement of a fourth in FY2020. Air Force plans call for procuring additional EC-37Bs at a rate of one per year until the planned total of 10 is reached.

Source: Artist rendering accompanying Brian W. Everstine, "Electronic Warfare Isn't Going Away—But EC130Hs Will," *Air Force Magazine*, February 2018.

Figure 4. EC-37B Compass Call Re-Host Aircraft (Artist's rendering).

[22] Modernization of EW systems on EC-130Hs is funded through a line item in the Modification of In-service Aircraft (i.e., Budget Activity 05) part of the Air Force's aircraft procurement account, in a line item entitled "Compass Call." In the Air Force's FY2020 budget submission, this is line item 53.

The Air Force's FY2020 budget submission projects that the first new EC-37B will enter the Air Force's inventory by the end of FY2020. The first two EC-37Bs are scheduled to achieve Initial Operational Capability (IOC) in 2023.[23]

L3 Technologies, a U.S. defense contractor involved in EW programs, is the prime contractor for the EC-37B. The EC-37B is based on the Gulfstream G550 commercial business jet, an aircraft that the Air Force also uses as the basis for its C-37B VIP transport aircraft. The Air Force states that EC-37Bs

> will receive Prime Mission Equipment (PME) from legacy donor EC-130H aircraft, as well as new, upgraded PME.... The re-hosted COMPASS CALL platform will utilize 70% of the PME off of the current airframe without modification; the remaining 30% of PME will be new or modified (repackaged) for the re-host. [Compared to the EC-130H,] the re-hosted COMPASS CALL aircraft will provide increased range, speed, endurance and operating altitude for better stand-off range and survivability. This will enable the USAF to effectively conduct Electronic Attack (EA) in an Anti-Access/Area Denial (A2AD) environment.[24]

The Air Force's acquisition strategy of replacing the EC-130H fleet by re-hosting their EW systems on new Gulfstream G550 aircraft was a subject of debate in Congress and contract-award protests.[25]

---

[23] Procurement of EC-37Bs is funded through the Other Aircraft (i.e., Budget Activity 04) part of the Air Force's aircraft procurement account, in a line item entitled "Compass Call." In the Air Force's FY2020 budget submission, this is line item 15. This line item is not to be confused with the previously discussed line item 53 in the Air Force's FY2020 aircraft procurement account, which is also called Compass Call. Line item 15 is for the procurement of new EC-37B aircraft, while line item 53 is for the procurement of Compass Call EW systems for both existing EC-130H aircraft and new EC-37B aircraft.

[24] *Department of Defense, Fiscal Year (FY) 2020 Budget Estimates, Air Force, Justification Book Volume 1 of 2 Aircraft Procurement, Air Force, Vol-1*, March 2019, p. 113.

[25] For press articles providing a review of the debate over the Compass Call replacement acquisition strategy, see Brian W. Everstine, "Electronic Warfare Isn't Going Away—But EC-130Hs Will," *Air Force Magazine*, February 2018; Rachel Cohen, "Congress Tells Air Force Not to Replicate EC-X Strategy for Other Programs," Inside Defense, November 17, 2017; Valerie Insinna, "L3 Gets Compass Call Contract, Names Gulfstrem As Airframe Provider," *Defense News*, September 8, 2017.

F-35 Joint Strike Fighter

The Lockheed Martin F-35 Joint Strike Fighter (**Figure 5**) is being procured in three versions for the Air Force (F-35A), Marine Corps (F-35B), and Navy (F-35C).

While the F-35's primary missions are air-to-ground combat (i.e., strike operations) and air-to-air combat (i.e., fighter operations), the F-35 has a built-in EW capability that is claimed by Lockheed Martin officials—the prime contractor manufacturing the aircraft—to be significantly greater than that of previous U.S. fighters and attack aircraft. Lockheed officials state that the F-35's EW system, designated AN/ASQ-239 serves as a signals collector system which provides: radar warning, identifies the geolocation of electronic emitters, tracks multiple aircraft simultaneously, provides high-gain (i.e., a highly focused radio antenna), high gain counter measures, and high gain electronic attack through the radar.[26] According to Lockheed officials these EW capabilities are designed to provide: wide-frequency coverage,[27] quick reaction time, high sensitivity and probability of intercept, accurate direction finding, track multiple aircraft, and provide self-protection countermeasures and jamming.

---

[26] Greg Lemons, Karen Carrington, Dr. Thomas Frey, and John Ledyard, Lockheed Martin Aeronautics Company, "F35 Mission Systems Design, Development, and Verification," AIAA Aviation Forum, June 25-29, Atlanta, GA, 2018 Aviation Technology, Integration, and Operations Conference, approved for public release 5/1/18, JSF18-502, p. 6, accessed April 2, 2019, at https://www.lockheedmartin.com/content/dam/lockheed-martin/eo/documents/webt/F35_Mission_Systems_Design_Development_and_Verification.pdf. Lockheed officials state on page 7 that "the EW system is common among the three F-35 variants, except for the forward Band 3/4 arrays, which employ longer elements for the F-35C Carrier Variant (CV). Also differing, the distance between the inboard and outboard arrays is less on the CV variant due to the wing fold. In addition to the EW Band 3/4 apertures, the radar MFA [multifunction] is employed to support EW functions."

[27] LtGen Steven Rudder has stated to the Senate Armed Services Committee that the F-35 can providing jamming across most frequencies, however there are deficiencies which can only be provided by the EA-18G. Statement of Lieutenant General Steven Rudder before the Senate Armed Services Committee Subcommittee on Seapower [hearing] on Navy and Marine Corps Aviation Programs in Review, March 6, 2018, pp. 57-58.

Source: Defense Visual Information Distribution Service (DVIDS), "National Cherry Festival Air Show 2018 [Image 2 of 12]," accessed April 1, 2019.

Figure 5. F-35 Joint Strike Fighter.

Lockheed Martin claims that due to the inherent, built-in electronic warfare capabilities the F-35 does not require a dedicated electronic attack aircraft to support it;[28] this would potentially free up other aircraft to perform electronic attack missions to protect less stealthy aircraft. To provide its organic jamming capability the F-35uses its active electronically scanned array (AESA) radar which teamed with advanced jamming algorithm packages, can potentially provide 10 times the jamming power of legacy aircraft.[29]

Figure 6 shows the location of EW system-related equipment on the F-35. During a 2018 hearing on the Navy and Marine Corps aviation program review, Lieutenant General Steven Rudder stated that although the Marine Corps was retiring the EA-6B, the Marine Corps' new F-35Bs

[28] Lockheed Martin, "Unprecedented Battlefield Access," accessed April 1, 2019. See also BAE Systems, "AN/ASQ239 F-35 electronic warfare / countermeasure system," accessed April 2, 2019, at https://www.baesystems.com/en-us/product/an-asq-239-f-35-ew-countermeasure-system.

[29] Tyler Rogoway, "Could the F-35 Become the Biggest Electronic Intelligence Collection System Ever Devised?" *The Drive*, July 6, 2016.

would have sufficient EW capability for most Marine Corps contingencies.[30]

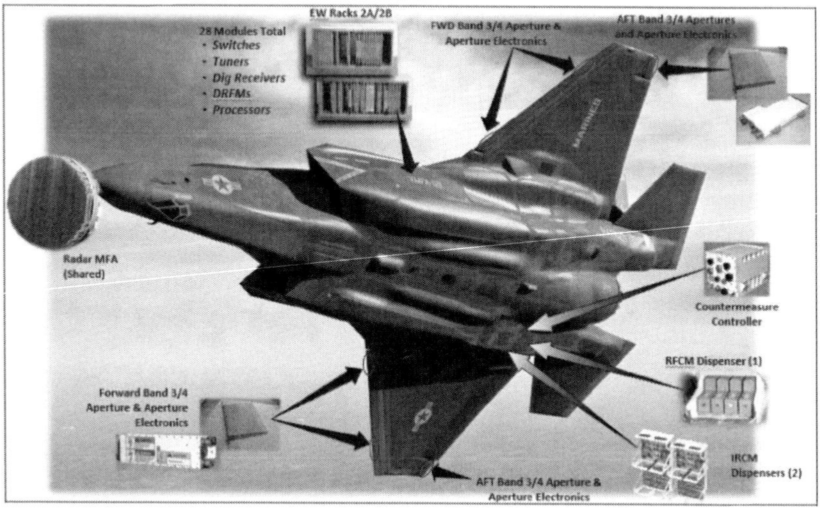

Source: Greg Lemons, Karen Carrington, Dr. Thomas Frey, and John Ledyard, Lockheed Martin Aeronautics Company, "F-35 Mission Systems Design, Development, and Verification," AIAA Aviation Forum, June 25-29, Atlanta, Georgia, 2018 Aviation Technology, Integration, and Operations Conference, approved for public release 5/1/18, JSF18-502, Figure 8, accessed April 2, 2019, at https://www.lockheedmartin.com/content/dam/lockheed-martin/eo/documents/webt/F35_Mission_Systems_Design_Development_and_Verification.pdf.

Figure 6. F-35 EW System-Related Equipment.

Airborne EW Payloads

DOD's primary airborne electronic attack payloads include the AN/ALQ-99 electronic attack suite, the Next Generation Jammer, and the Miniature Air Launched Decoy-Jammer. Each of these systems is discussed briefly below.

---

[30] Statement of Lieutenant General Steven Rudder before the Senate Armed Services Committee Subcommittee on Seapower [hearing] on Navy and Marine Corps Aviation Programs in Review, March 6, 2018, pp. 57-58.

AN/ALQ-99 Tactical Jamming System (TJS)

The AN/ALQ-99 tactical jamming system (Figure 7 and Figure 8; see also Figure 2) consists of a series of electronic jamming pods. The system was originally developed in the 1970s for the EA-6B, and it was also used by the EF-111A. The system has been updated over time and is currently carried by EA/18Gs. The current version of the system, called the ALQ-99F(V), achieved IOC in 1999. Navy plans call for replacing the ALQ-99 with the Next Generation Jammer (see next section). The Navy states that the ALQ-99 "is the only airborne tactical jamming system in the Department of Defense inventory. [The] ALQ-99 [system] is facing material and technological obsolescence and cannot counter all current, much less future, threats."[31]

Source: Annotated U.S. Navy photo as printed in Government Accountability Office, *Next Generation Jammer[:] DOD Should Continue to Assess Potential Duplication and Overlap As Program Moves Forward*, GAO-13-642, August 2013, p. 5.

Figure 7. EA-18G Growler Equipped with AN/ALQ-99F Tactical Jamming System.

[31] Department of the Navy, *U.S. Navy Program Guide 2017*, p. 27. The Navy's use of the word "tactical" refers to how the system is carried on a tactical fighter aircraft, as opposed to an aircraft like the EC-130H Compass Call.

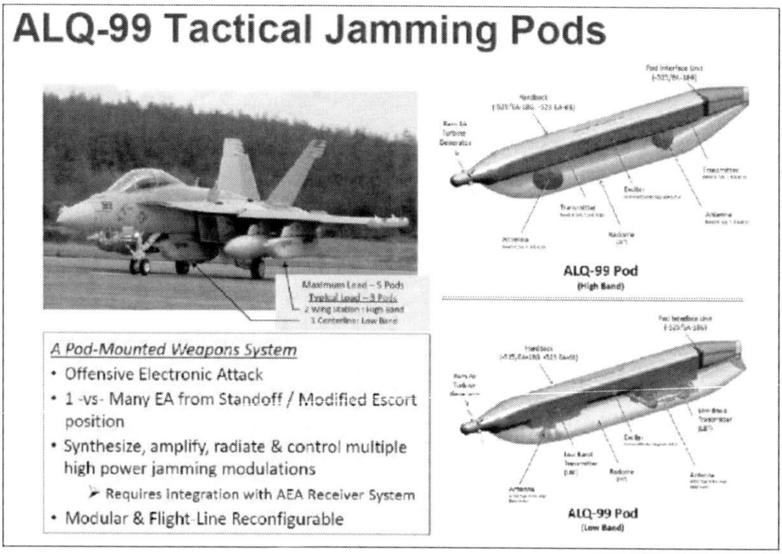

Source: Briefing slide accompanying Joseph Trevithick and Tyler Rogoway, "Navy's Next Gen Jammer Is Three Pods, Not One, And Competition For One Of Them Just Heated Up," *The Drive*, October 25, 2018. In this press report, the caption for the briefing slide states: "An RAAF [Royal Australian Air Force] briefing slide showing the interior layout of the two types of AN/ALQ-99 pod."

Figure 8. ALQ-99 Tactical Jamming Pods.

Next Generation Jammer (NGJ)

As mentioned above, Navy plans call for replacing the ALQ-99 with a new EW system called the Next Generation Jammer (NGJ) (Figure 9, Figure 10, and Figure 11).

The NGJ is being developed in three increments designed to jam across three radio frequency bands to prevent adversaries from using their communications and radar systems.[32] The first increment, which is to provide EW capability in mid-band frequencies, was previously referred to

---

[32] Radars and radios operate at different frequencies depending on what mission they are trying to accomplish. Lower frequencies (i.e., "low-band") provide a longer range picture of the airspace, but cannot be used to launch a missile. Middle-band radars can provide target quality pictures for missiles, however do not have as long of a range. High-band radars can provide high-fidelity pictures, however they shorter ranges and the radio frequencies can be disrupted by objects and even water molecules. For more information see CRS In Focus IF11155, *Defense Primer: Military Use of the Electromagnetic Spectrum*, by John R. Hoehn.

as Increment 1, but is now called the AN/ALQ-249 system or the Next Generation Jammer—Mid Band (NGJ-MB). The next increment, which is to provide EW capability in low-band frequencies, was previously referred to as Increment 2, is now called the Next Generation Jammer—Low Band (NGJ-LB). The remaining increment, currently called Increment 3, is to provide EW capability in high-band frequencies. DOD states that "the order of development [of the increments] was determined by the assessed capabilities of the developing threat and shortfalls of the legacy system to counter those capabilities, with Inc 1 [the Increment 1 system] covering the most critical threats."[33]

Source: Photo accompanying Sam LaGrone, "L3, Northrop Selected for Next Generation Jammer Work; Program Stalled After Raytheon Protest," *USNI News*, August 21, 2018.

Figure 9. Next Generation Jammer—Mid Band (NGJ-MB) (Artist's rendering).

Raytheon was awarded the contract for developing NGJ-MB. L3 Systems, Northrop Grumman, and Harris were awarded the contract for

[33] Department of the Navy, *U.S. Navy Program Guide 2017*, p. 27 and Department of Defense, *Director, Operational Test & Evaluation, FY 2016 Annual Report*, December 2016, p. 303.

developing NGJ-LB.[34] The Navy's FY2020 budget submission states that NGJ-MB is scheduled to achieve IOC in the fourth quarter of FY2022.[35]

Source: Department of Defense, Director, Operational Test & Evaluation, FY 2016 Annual Report, December 2016, p. 303.

Figure 10. Next Generation Jammer—Mid Band (NGJ-MB) (Artist's rendering).

The NGJ program has been a subject of congressional oversight for several years.[36] An August 2018 press report states the Navy stalled in its development of the NGJ, however with the renewed focus of "great power competition," particularly with Russia and China, the NGJ has been given

[34] See, for example, Joseph Trevithick and Tyler Rogoway, "Navy's Next Gen Jammer Is Three Pods, Not One, And Competition For One Of Them Just Heated Up," *The Drive*, October 25, 2018; Sam LaGrone, "L3, Northrop Selected for Next Generation Jammer Work; Program Stalled After Raytheon Protest," *USNI News*, August 21, 2018.

[35] See also Julian Kerr, "Avalon [Airshow] 2019: Next Generation Jammer on Track for US Navy IOC in 2022," *Jane's Defence Weekly*, February 28, 2019; Paul McLeary, "Navy's New EW Jammer to be Ready in 2022, *Breaking Defense*, July 17, 2018.

[36] See, for example, Government Account Ability Office, *Weapon Systems Annual Assessment[:] Knowledge Gaps Pose Risks to Sustaining Recent Positive Trends*, GAO-18-360SP, April 2018, pp. 102-103; Department of Defense, *Director, Operational Test & Evaluation, FY 2016 Annual Report*, December 2016, pp. 303-304; Government Accountability Office, *Next Generation Jammer[:] DOD Should Continue to Assess Potential Duplication and Overlap As Program Moves Forward*, GAO-13-642, August 2013, 27 pp.

increased importance and priority.[37] Frank Kendall, when he served as the Undersecretary of Acquisition, Technology and Logistics decided to accelerate the program.

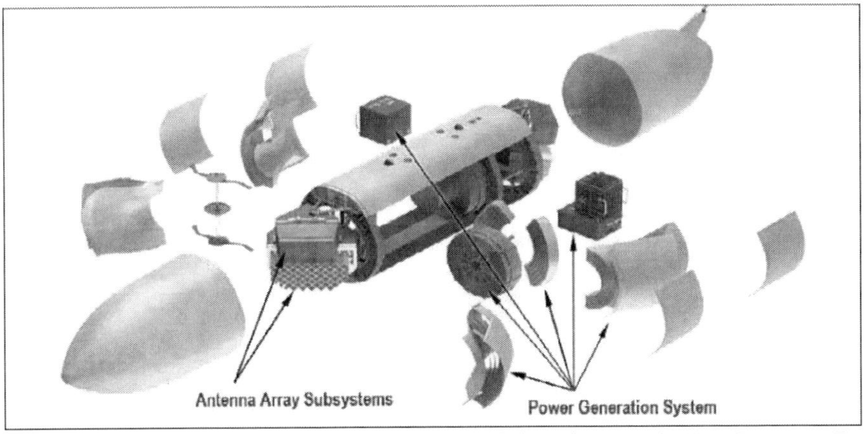

Source: Illustration accompanying Joseph Trevithick, "The Navy's New Jammers For Its EA-18G Growlers Cut Back Their Range More Than The Old Pods," *The Drive*, November 28, 2018.

Figure 11. Next Generation Jammer—Mid Band (NGJ-MB).

The Navy's FY2020 budget submission requests $524.3 million for PE 0604274N in FY2020. The budget submission projects annual funding to decline in subsequent years, to $178.4 million in FY2022 and zero funding thereafter as research and development work on NGJ-MB is completed and NGJ-MB transitions from research and development to procurement.[38] The budget submission estimates the total research and development cost of NGJ-MB at $3,985.0 million (i.e., about $4.0 billion), of which $2,848.2 million (i.e., about $2.8 billion) has been received through FY2019.

[37] Sam LaGrone, "L3, Northrop Selected for Next Generation Jammer Work; Program Stalled After Raytheon Protest," *USNI News*, August 21, 2018.

[38] Research and development work on NGJ-MB is funded in Program Element (PE) 0604274N, Next Generation Jammer (NGJ)—a line item in the Navy's research and development account. In the Navy's FY2020 budget submission, PE 0604274N is line 116 in the account.

The Navy's FY2020 budget submission requests $6.2 million for PE0604274N—the first procurement funding requested for NGJ-MB.[39] The submission projects that in subsequent years, as procurement of NGJ-MB ramps up, annual funding for this line item would increase to $144.7 million in FY2021 and $534.1 million by FY2024. The submission estimates the total procurement cost of NGJ-MB at $4,830.9 million (i.e., about $4.8 billion).

The Navy's FY2020 budget submission requests $111.1 million for this PE in FY2020 and projects annual funding to increase in subsequent years, to $241.5 million in FY2024.[40] The submission estimates the total research and development cost of NGJ-LB at $3,499.1 (i.e., about $3.5 billion), of which $178.3 million has been received through FY2019.

Impact of Next Generation Jammer on Range of EA-18G

Because the NGJ reportedly produces more drag on the EA-18G according to the Navy the Next Generation Jammer has the potential of reducing the operational range of the EA-18G. A

November 28, 2018, press report states that the NGJ Mid-band pod produces more drag than the current ALQ-99.[41] Raytheon's proposal for the low-band pod was partially rejected as a result of increased drag over competing designs. These increases in drag have been reported to reduce the operational range of the EA-18G. The specific impact on range is classified.[42]

Miniature Air-Launched Decoy (MALD) and Jammer (MALD-J)

DOD states that the Miniature Air Launched Decoy (MALD) and Jammer (MALD-J), also designated ADM-160 (Figure 12 and Figure 13),

---

[39] Procurement of NGJ-MB is funded through a line item called Next Generation Jammer (NGJ) in the Modification of Aircraft (i.e., Budget Activity 05) part of the Navy's aircraft procurement account. In the Navy's FY2020 budget submission, it is line 59 in the account.

[40] Research and development work on NGJ-LB is funded in PE 0604282N, Next Generation Jammer (NGJ) Increment II, another line item in the Navy's research and development account. In the Navy's FY2020 budget submission, this PE is line 118 in the account.

[41] Joseph Trevithick, "The Navy's New Jammers For Its EA-18G Growlers Cut Back Their Range More Than The Old Pods," *The Drive*, November 28, 2018.

[42] The Raytheon Company, File B-416578.1, October 22, 2018. Accessed at: https://www.gao.gov/docket/B-416578.1.

is designed as a low-cost, expendable vehicle that can replicate the flight and radar signatures of manned aircraft. The MALD-J adds an electronic attack component. According to the DOD "MALD-J is designed to support an airborne strike force to achieve mission success by jamming enemy radars and air-defense systems by degrading/denying detection of friendly aircraft or munitions."[43]

MALD has a reported range of about 500 nautical miles.[44] It was first developed in the mid-1990s, and more than 2,000 have been produced. A new version, designated MALD-X, is now being developed. An August 2018 press report states that

> MALD-X enhances the modular nature of the mini cruise missile with the ability to accommodate different electronic warfare payloads that are more advanced than those found on MALD-J. What is planned to come out of MALD-X is a networked decoy that can use its adaptive electronic warfare payload to deliver electronic attacks on air defense nodes autonomously or at the direction of operators from a afar in a semi-autonomous fashion.[45]

A derivative of MALD-J and MALD-X, designated MALD-N, is being developed for use on Navy F/A-18E/F strike fighters.[46]

---

[43] Department of Defense, *Director, Operational Test and Evaluation, FY 2017 Annual Report*, January 2018, p. 265.

[44] See, for example, Raytheon, "MALD Decoy, Disrupting Enemy Air Defense Systems," undated, accessed April 4, 2019.

[45] Tyler Rogoway, "Recent MALD-X Advanced Air Launched Decoy Test Is A Much Bigger Deal Than It Sounds Like," *The Drive*, August 24, 2018. See also Colin Clark, "Did We Put This Bug In SCO's Ear? MALD-X Tested For Swarm EW," *Breaking Defense*, August 23, 2018; Kimberly Underwood, "U.S. Military Takes MALD-X Outfitted with Electronic Warfare to the Air," *Signal*, August 24, 2018.

[46] Matthew Nelson, "Raytheon to Develop Miniature Aerial Decoy Tech for Navy," *Executive Biz*, January 31, 2019; Richard Scott, "MALD-N Programme Transitions into TMRR Phase, Splits off Jammer Payload," *Jane's Missiles & Rockets*, October 2, 2018; Jason Sherman, "Navy Taps Raytheon to Launch Next-Generation EW Decoy: the MALDN," *Inside Defense*, September 26, 2018; Tamir Eshel, "US Navy to Field a Networked Decoy/Jammer," *Defense Update*, August 26, 2018; Tyler Rogoway, "Recent MALD-X Advanced Air Launched Decoy Test Is A Much Bigger Deal Than It Sounds Like," *The Drive*, August 24, 2018.

Source: Photo accompanying Tyler Rogoway, "Recent MALD-X Advanced Air Launched Decoy Test Is A Much Bigger Deal Than It Sounds Like," *The Drive*, August 24, 2018. The caption to the photo states: "MALDs loaded up on a B-52H."

Figure 12. Miniature Air-Launched Decoy (MALD) (Shown fitted to a B-52H bomber).

Source: Photo accompanying Tyler Rogoway, "Recent MALD-X Advanced Air Launched Decoy Test Is A Much Bigger Deal Than It Sounds Like," The Drive, August 24, 2018.

Figure 13. Miniature Air-Launched Decoy (MALD).

Potential Issues for Congress

Congressional EW Working Group (EWWG)

Given their interest in, and concerns about, U.S. EW capabilities in the era of renewed great power competition, some Members of Congress have met in recent years through the Electronic Warfare Working Group (EWWG).[47] In the 115th Congress, Representative Bacon introduced the Joint Electromagnetic Spectrum Operations Readiness Act of 2018 (H.R. 5522). This bill would have asked the DOD to

- develop a joint campaign modeling capability to model electromagnetic spectrum effect on operations,
- assess capabilities and capacities of EW platforms associated with operational plans, and
- develop an interim and annual report on programs and personnel assigned to EW missions.[48]

An identical bill in the Senate was referred to the Senate Armed Services Committee.[49]

---

[47] See, for example, "EWWG Briefing: "Electronic Warfare & Electromagnetic Spectrum Operations," concerning a briefing scheduled for October 3, 2017, accessed March 11, 2019, at http://dearcolleague.us/2017/09/ewwg-briefingelectronic-warfare-electromagnetic-spectrum-operations/; "AOC [Association of Old Crows] Advocacy: EW Working Group Hosts EW/EMSO Educational Briefing Series for Congress," March 29, 2017, accessed March 11, 2019, at http://www.ecrow.org/articles/index-v4.asp?aid=432979&issueID=52139; Association of Old Crows, invitation to panel discussion on the electromagnetic spectrum scheduled for June 21, 2016, accessed March 11, 2019, at http://www.ecrow.org/pdf/EWWG_invitation_6.21.pdf; Association of Old Crows, photo of June 21, 2016, meeting, accessed March 11, 1019, at https://www.facebook.com/aocrows/photos/electronic-warfare-working-group-ewwghosts-electromagnetic-spectrum-ems-panel-w/1056968031018192/. The Association of Old Crows (AOC) is a professional association for those with an interest in EW. For more on the AOC, see Association of Old Crows, "Mission & History," accessed March 11, 2019, at https://www.crows.org/page/mission andhistory.

[48] H.R. 5522.

[49] S. 2949.

Airborne EW as a DOD Priority

One potential oversight issue for Congress is whether DOD is giving too little, too much, or the right amount of priority to airborne EW programs in its planning and budgeting relative to other U.S. military EW programs (such as those for U.S. ground forces or Navy surface ships) and to other DOD non-EW priorities, particularly in the context of renewed great power competition and improvements in air defense and EW capabilities by Russia, China, and other potential adversaries.

Congress may consider developments such as Russia's deployment and sales to other countries of advanced air defense systems. Some observers have expressed concern about Russia's ability to use its advanced air defense systems, such as its S-400 surface-to-air (SAM) missile system, to establish hard-to-penetrate anti-access/area-denial (A2/AD) zones around defended areas in Europe and Middle East, and for countries that purchase Russian-made air defense systems, such as China, to do something similar in other regions.[50] Other observers state that the capabilities of Russia's A2/AD air defense systems have been overrated.[51] The Defense Intelligence Agency states that Russian air defense are among the best in the world, and they continue to develop highly-capable systems which they export to countries like China interested in acquiring long range defensive systems.[52]

---

[50] See, for example, Michael Cecire, "NATO Is Focusing on the Wrong Russian Threat in Eastern Europe," *World Politics Review*, October 26, 2018; "China's Anti-Access Area Denial," Missile Defense Advocacy Alliance, August 24, 2018; Dominik Jankowski, "Six Ways NATO Can Address the Russian Challenge," Atlantic Council, July 4, 2018; "The Baltic Dilemma: NATO Security in the Russian Anti-Access Bubble," Royal Danish Defence College, May 3, 2018 (announcement for a conference scheduled for May 29, 2018); Ian Williams, The Russia – NATO A2AD Environment," Center for Strategic and International Studies (Missile Threat, CSIS Missile Defense Project), January 3, 2017.

[51] See, for example, Robert Dalsjö, Christofer Berglund, and Michael Jonsson, *Bursting the Bubble, Russian A2/AD in the Baltic Sea Region: Capabilities, Countermeasures, and Implications*, FOI, Swedish Defence Research Agency, March 2019, 114 pp.; David Axe, "A New Report Claims Russia's Mighty Missiles Might Not Be So Mighty After-all," National Interest, March 7, 2019. See also Jyri Raitasalo, "It Is Time to Burst the Western A2/AD Bubble," Kungl Krigsvetenskapsakademien (Royal Swedish Academy of War Sciences), June 16, 2017.

[52] Defense Intelligence Agency, *Russia Military Power, Building a Military to Support Great Power Aspirations*, 2017, p. 79.

DOD states that China's air force has one of the largest air defense forces, with a series of advanced long-range surface-to-air missiles. These consist of the Russian S-300PMU, the domestically produced CSA-9, and the recently fielded Russian S-400 system.[53]

An April 2018 press report stated (emphasis added):

> One of the "wicked problems" [U.S.] commandos are facing now is in Syria, which [U.S. Army General Tony] Thomas called the "most aggressive electronic warfare environment on the planet from our adversaries. They are testing us every day, knocking our communications down, disabling our EC-130s, etc."[54]

Another factor Congress may consider is how, in a situation of finite DOD funding, devoting more funding to airborne EW programs would affect funding for other EW priorities, or DOD non-EW priorities, and what the resulting net change would be in overall U.S. military capabilities.

Mix of Airborne EW Capabilities and Investments

Another potential oversight issue for Congress is whether DOD's proposed mix of airborne EW capabilities and investments is appropriate given the current and projected capabilities of potential adversaries such as Russia and China. Specifically, what is DOD's vision by combining specialized tactical EW aircraft such as the EA-18G, standoff EW aircraft such as the EC-130H and EC-37B, strike fighters with embedded EW capabilities such as the F-35, and air-launched decoys and jammers with growing numbers of stealthy fifth-generation F-35s? What evolutions are occurring in U.S. military operational concepts? Is it appropriate, for example, that the Air Force and Marine Corps no longer operate their own specialized tactical EW aircraft, while the Navy continues to operate and invest in the EA-18G and its Next Generation Jammer EW pods? More

---

[53] Department of Defense, Annual Report to Congress: Military and Security Developments Involving the People's Republic of China 2018, pp. 34-35.

[54] Howard Altman, "SOCom Leader Wanted to Toss Google Exec from Car. Because He Was Right." *Tampa Bay Times*, April 26 (updated April 27), 2018.

generally, to what degree do the airborne EW capabilities of the Air Force, Navy, and Marine Corps overlap, and is that overlap appropriate?

Role of Emerging Technologies

A third potential issue for Congress is how DOD uses advances in technology. Electronic attack platforms have evolved from the manned platforms with relatively large crew sizes, such as the EA-6B Prowler and the EC-130H Compass Call, to the EA-8G Growler with a crew of two and the MALD-J, which does not have crew and is a standoff weapon. Evolving A2/AD environments potentially make traditional stand-in jamming too dangerous for manned aircraft. Therefore, Congress may consider policy for DOD regarding developing platforms that are capable of operating in A2/AD environments. What unmanned EW programs does DOD currently fund? Does DOD plan to develop additional stand-in jamming systems?

EC-37B Compass Call Re-Host Aircraft Procurement

Another oversight issue for Congress concerns the Air Force's planned quantity and rate for procuring EC-37Bs and replacing EC-130Hs. The Air Force plans to replace 14 EC-130H aircraft with 10 EC-37Bs. The Air Force currently maintains 10 EC-130H Compass Calls for operations, with one aircraft devoted to testing and three additional aircraft in back-up inventory. How does the Air Force plan to use a smaller fleet of aircraft? Would 10 EC-37B aircraft be able to meet operational demands?

Some Members of Congress have expressed an interest in procuring EC-37Bs more quickly than the Air Force plans, so as to accelerate the replacement of EC-130Hs with EC-37Bs. The committee and conference report language bearing on this issue for the John S. McCain National Defense Authorization Act for Fiscal Year 2019 and the FY2019 DOD appropriations act appear below.

APPENDIX. RECENT CONGRESSIONAL ACTION

John S. McCain National Defense Authorization Act for Fiscal Year 2019 (H.R. 5515/S. 2987 /P.L. 115-232)

The House Armed Services Committee, in its report (H.Rept. 115-676 of May 15, 2018) on H.R. 5515, the FY2019 National Defense Authorization Act, stated that

> [t]he committee supports the Air Force's efforts to recapitalize the aging EC–130H Compass Call fleet with the more capable EC–37 type aircraft. The committee notes that the Air Force must first comply with the National Defense Authorization Act for Fiscal Year 2017 (Public Law 114–328) and the National Defense Authorization Act for Fiscal Year 2018 (Public Law 115–91) before it can carry on with the transition plan. The Air Force requested $108.1 million for fiscal year 2019 for one EC–37. The committee is concerned that the Air Force plan to procure one aircraft per year over 10 years in order to recapitalize this fleet is not the most efficient way to move the capability to the field quickly, and may put the Compass Call mission at unacceptable risk of mission failure. Therefore, the committee directs the Secretary of the Air Force to provide a briefing to the House Committee on Armed Services by February 1, 2019, on the Compass Call transition plan. This plan should include:
>
> (1) courses of action to accelerate the recapitalization of the EC–130H fleet and Baseline 4 development and deployment for incoming EC–37 aircraft;
> (2) attendant timelines for each course of action;
> (3) cost estimates for each course of action;
> (4) recommended course of action and a plan to manage both fleets while supporting combatant commander requirements; and
> (5) an assessment of the potential for future cooperative development and procurement of EC–37B Compass Call aircraft by the Royal Air Force of the United Kingdom and the Royal Australian Air Force in a way the leverages the best practices of the RC–135 cooperative program arrangement with the Royal Air Force of the United Kingdom.

The Senate Armed Services Committee, in its report (S.Rept. 115-262 of June 5, 2018) on S. 2987, the John S. McCain National Defense Authorization Act for Fiscal Year 2019, stated the following:

> The committee supports the Air Force's efforts to recapitalize the aging EC–130H Compass Call fleet with the EC–37 type aircraft. The committee notes that before it can carry on with the transition plan, the Department of Defense must first comply with the related provisions in the National Defense Authorization Act for Fiscal Year 2017 (Public Law 114–328) and the National Defense Authorization Act for Fiscal Year 2018 (Public Law 115–91). While the committee notes that Department has submitted the certification required by the National Defense Authorization Act for Fiscal Year 2018 (Public Law 115–91), delays in satisfying the requirement has led to a work stoppage on the program lasting at least six weeks. The committee is concerned about the potential for further work stoppages should the Secretary of the Air Force fail to make a timely determination that the EC–37B has a high likelihood of meeting combatant requirements, as required by the National Defense Authorization Act for Fiscal Year 2017 (Public Law 114–328). The committee encourages the Secretary of the Air Force to make a timely determination for this requirement to avoid further program delays and cost overruns.
>
> Therefore, the committee directs the Secretary of the Air Force, not more than 60 days after the determination required by the National Defense Authorization Act for Fiscal Year 2017 (Public Law 114–328) is made, to provide a briefing to the congressional defense committees on the Compass Call transition plan. This plan should include:
>
> (1) Courses of action to accelerate the recapitalization of the EC–130H fleet and Baseline 4 development and deployment for incoming EC–37 aircraft;
> (a) attendant timelines for each course of action;
> (b) cost estimates for each course of action; and
>
> (2) Recommended course of action and a plan to manage both fleets while supporting combatant commander requirements.

FY2019 DOD Appropriations Act (Division A of H.R. 6157/P.L. 115-245)

The House Appropriations Committee, in its report (H.Rept. 115-769 of June 20, 2018) on the FY2019 DOD appropriations act (H.R. 6157), stated the following:

> The [committee's] recommendation [of FY2019 procurement funding for the program] includes an increase of $194,000,000 above the budget request to procure and modify one additional EC–37B Compass Call aircraft, with the expectation that such funds will allow the Air Force to accelerate the fielding of the fourth such aircraft to meet combatant commander needs and mitigate performance concerns regarding the legacy EC–130H fleet. The Committee recommends that the Secretary of the Air Force consider increasing the procurement of EC–37B aircraft to two per year if such a pace of recapitalization can be achieved without unduly disrupting the operational availability of Compass Call capability for the combatant commanders. (Page 188)

In final action on the FY2019 DOD Appropriations Act (Division A of H.R. 6157/P.L. 115-245 of September 28, 2018), Congress increased the requested amount for procurement of new EC-37Bs by $108 million for "Program increase - accelerate fourth EC-37B aircraft."[55]

[55] Joint Explanatory Statement for H.R. 6157, PDF page 203 of 559, line 15.

In: Key Congressional Reports for May 2019 ISBN: 978-1-53616-335-3
Editor: Aston Dunkley

Chapter 5

THE POTENTIAL DECLINE OF CASH USAGE AND RELATED IMPLICATIONS*

David W. Perkins

ABSTRACT

Electronic forms of payment have become increasingly available, convenient, and cost efficient due to technological advances in digitization and data processing. Anecdotal reporting and certain analyses suggest that businesses and consumers are increasingly eschewing cash payments in favor of electronic payment methods. Such trends have led analysts and policymakers to examine the possibility that the use and acceptance of cash will significantly decline in coming years and to consider the effects of such an evolution.

Cash is still a common and widely accepted payment system in the United States. Cash's advantages include its simplicity and robustness as a payment system that requires no ancillary technologies. In addition, it provides privacy in transactions and protection from cyber threats or financial institution failures. However, using cash involves costs to businesses and consumers who pay fees to obtain, manage, and protect cash and exposes its users to loss through misplacement, theft, or

* This is an edited, reformatted and augmented version of a Congressional Research Service publication R45716, prepared for Members and Committees of Congress dated May 10, 2019.

accidental destruction of physical currency. Cash also concurrently generates government revenues through "profits" earned by producing it and by acting as interest-free liabilities to the Federal Reserve (in contrast to reserve balances on which the Federal Reserve pays interest), while reducing government revenues by facilitating some tax avoidance.

The relative advantages and costs of various payment methods will largely determine whether and to what degree electronic payment systems will displace cash. Traditional noncash payment systems (such as credit and debit cards and interbank clearing systems) involving intermediaries such as banks and central banks address some of the shortcomings of cash payments. These systems can execute payments over physical distance, allow businesses and consumers to avoid some of the costs and risks of using cash, and are run by generally trusted and closely regulated intermediaries. However, the maintenance and operation of legacy noncash systems involve their own costs, and the intermediaries charge fees to recoup those costs and earn profits. The time it takes to finalize certain transactions—including crediting customer accounts for check or electronic deposits—can lead to consumers incurring additional costs. In addition, these systems involve cybersecurity risks and generally require customers to divulge their private personal information to gain system access, which raises privacy concerns.

To date, the migration away from cash has largely been in favor of traditional noncash payment systems; however, some observers predict new alternative systems will play a larger role in the future. Such alternative systems aim to address some of the inefficiencies and risks of traditional noncash systems, but face obstacles to achieving that aim and involve costs of their own. Private systems using distributed ledger technology, such as cryptocurrencies, may not serve the main functions of money well and face challenges to widespread acceptance and technological scalability. These systems also raise concerns among certain observers related to whether these systems could facilitate crime, provide inadequate protections to consumers, and may adversely affect governments' ability to implement or transmit monetary policy. The potential for increased payment efficiency from these systems is promising enough that certain central banks have investigated the possibility of issuing government-backed, electronic-only currencies—called central bank digital currencies (CBDCs)—in such a way that the benefits of certain alternative payment systems could be realized with appropriately mitigated risk. How CBDCs would be created and function are still matters of speculation at this time, and the possibility of their introduction raises questions about the appropriate role of a central bank in the financial system and the economy.

If the relative benefits and costs of cash and the various other payment methods evolve in such a way that cash is significantly displaced as a commonly accepted form of payment, that evolution could have a number of effects, both positive and negative, on the economy and society. Proponents of reducing cash usage (or even eliminating it all together and becoming a *cashless society*) argue that doing so will generate important benefits, including potentially improved efficiency of the payment system, a reduction of crime, and less constrained monetary policy. Proponents of maintaining cash as a payment option argue that significant reductions in cash usage and acceptance would further marginalize people with limited access to the financial system, increase the financial system's vulnerability to cyberattack, and reduce personal privacy. Based on their assessment of the magnitude of these benefits and costs and the likelihood that market forces will displace cash as a payment system, policymakers may choose to encourage or discourage this trend.

THE ONCE, BUT PERHAPS NOT FUTURE, KING

Because of technological advances in digitization and data processing, electronic forms of payment have become increasingly available, convenient, and cost efficient. Established technologies, such as credit and debit cards, have long been a popular payment option. In addition, new payment methods (e.g., PayPal's Venmo app and Square's point-of-sale hardware, among others) use underlying traditional banking and payments systems to make electronic payments less expensive and more available to individuals and small businesses. Newer digital currencies, such as cryptocurrencies, offer alternative (though not yet widely adopted) options that have a high degree of independence from traditional systems.[1]

Although cash remains an important method of payment in the United States (see Figure 1), anecdotal reporting suggests that various electronic payment systems have become so effective and inexpensive relative to cash payments that some U.S. businesses—even those at which sales generally have a low dollar value—are increasingly choosing not to accept

[1] Cryptocurrencies and virtual currencies are alternative, often decentralized, electronic payment systems using money that generally is not backed by government fiat. The "Private Payment Systems Using Distributed Ledgers" section in this chapter describes them in more detail. For a more detailed examination of cryptocurrencies, see CRS Report R45427, *Cryptocurrency: The Economics of Money and Selected Policy Issues*, by David W. Perkins.

cash.[2] In some developed countries, such as Sweden, cash payments are becoming relatively scarce.[3] In addition, a number of central banks worldwide are examining the possibility of issuing government-backed digital currencies that exist only electronically.[4] These trends suggest that due to buyer or seller preference or government policy, the role of cash in the payment system may continue to decline, perhaps significantly, in coming years.[5]

Some observers have examined the consequences of an evolution away from cash.[6] Proponents of reducing the use of physical currency (or even eliminating it all together and becoming a *cashless society*) argue that it will generate important benefits, including potentially improved efficiency of the payment system, a reduction of crime, and less constrained monetary policy.[7] Proponents of maintaining cash as a payment option argue that significant reductions in cash usage and acceptance would further marginalize people with limited access to the financial system, increase the financial system's vulnerability to cyberattack, and reduce personal privacy.[8] Given developments and debate in this area, Congress may consider policy issues related to the declining use of cash relative to electronic forms of payment.

---

[2] For example, see Andy Newman, "Cash Might Be King, But They Don't Care," *The New York Times*, December 25, 2017, at https://www.nytimes.com/2017/12/25/nyregion/no-cash-money-cashless-credit-debit-card.html.

[3] Maddy Savage, "Why Sweden Is Close to Becoming a Cashless Economy," *BBC News*, September 12, 2017, at https://www.bbc.com/news/business-41095004.

[4] Central bank digital currencies are a potential new form of digital central bank money that is different from reserves or settlement balances held by commercial banks at central banks. The "Central Bank Digital Currencies" section in this chapter describes them in more detail.

[5] Speculating on the role each of these forces may play in a potential future decline in cash usage or acceptance is beyond the scope of this chapter. Instead, two international case studies are discussed in the **Appendix** of this chapter.

[6] For example, see Kenneth S. Rogoff, *Costs and Benefits to Phasing Out Paper Currency*, National Bureau of Economic Research, Working Paper no. 20126, May 2014, at https://www.nber.org/papers/w20126.pdf. Hereinafter Kenneth S. Rogoff, *Costs and Benefits to Phasing Out Paper Currency*.

[7] Kenneth S. Rogoff, *Costs and Benefits to Phasing Out Paper Currency*, pp. 1-6, 10-11.

[8] James J. McAndrews, "Should We Move to a Mostly Cashless Society?" *Wall Street Journal*, September 24, 2017, at https://www.wsj.com/articles/should-we-move-to-a-mostly-cashless-society-1506305220, and Laurens Cerulus and Cat Contiguglia, "Central Bankers Warn of Chaos in a Cashless Society," *Politico*, August 14, 2018, at https://www.politico.eu/article/central-bankers-fear-cybersecurity-chaos-in-a-cashless-society/.

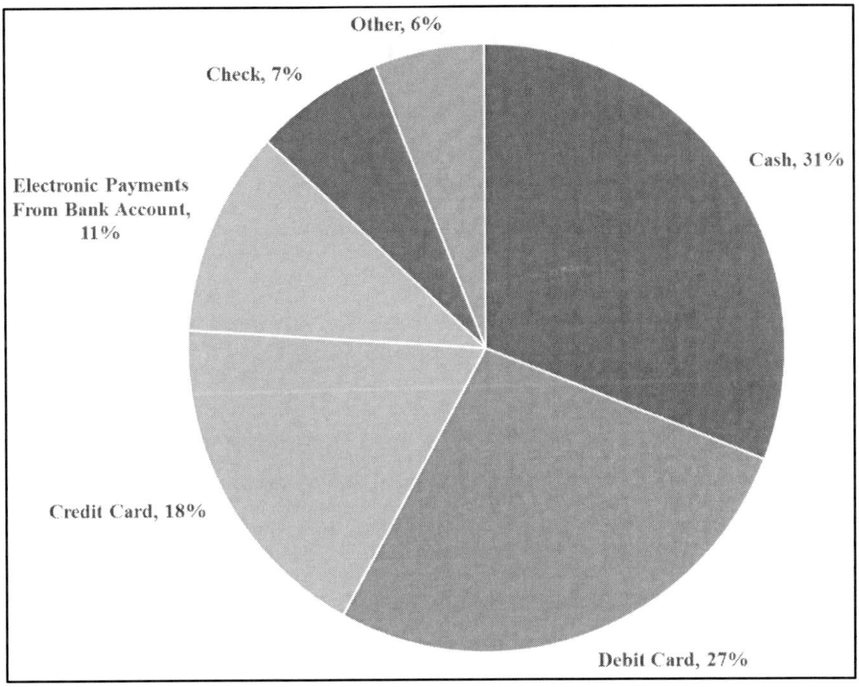

Source: Shaun O'Brien, *Understanding Consumer Cash Use: Preliminary Findings from the 2016 Diary of Consumer Payment Choice*, Federal Reserve Bank of San Francisco, Fednotes, November 28, 2017.

Figure 1. Payments Type, Share of All Transactions.

This chapter is divided into two parts. The first part analyzes cash and noncash payment systems, and the second analyzes potential outcomes if cash were to be significantly displaced as a commonly accepted form of payment. Part I describes the characteristics of cash and the various electronic payment systems that could potentially supplant cash. The noncash payment systems include traditional electronic payment systems (such as credit cards or payment apps) and alternative electronic payment systems, focusing on private systems using distributed ledger technology (such as cryptocurrencies) and central bank digital currencies (which are only under consideration by central banks at this time). Part I also examines the advantages and costs specific to each payment system and the potential obstacles to the adoption of alternative electronic payment systems. Part II of this chapter analyzes the potential implications of a reduced role of cash payments in the economy, including potential benefits, costs, and risks.[9] The report also includes an Appendix that presents two international case studies of economies in which noncash payment systems rapidly expanded.

PART I: ANALYSIS OF CASH AND NONCASH PAYMENT SYSTEMS

This section provides analysis of cash, traditional noncash payment systems, and potential alternative payment systems. It describes the characteristics, presents usage data, and analyzes the advantages and costs of each system. It also includes a discussion on the potential decline in cash usage and a short inset on the legality of businesses' refusing to accept cash.

Physical Currency—Cash

Overview

How well something serves as money in a payment system depends on how well it serves as (1) a medium of exchange, (2) a unit of account, and (3) a store of value. To function as a *medium of exchange*, the thing must be tradable and agreed to have value. To function as a *unit of account*, the thing must act as a good measurement system. To function as a *store of*

[9] A detailed examination of the economic function of money itself is beyond the scope of this chapter. For more information on that topic, see CRS Report R45427, *Cryptocurrency: The Economics of Money and Selected Policy Issues*, by David W. Perkins. An examination of how money maintains its value through Federal Reserve monetary policy is also beyond the scope of this chapter. For more information on that topic, see CRS Report RL30354, *Monetary Policy and the Federal Reserve: Current Policy and Conditions*, by Marc Labonte.

value, the thing must be able to purchase approximately the same value of goods and services at some future date as it can purchase now.[10]

Currently, cash continues to serve the three functions of money well as part of a robust, physical payment system. Physical currency can be carried easily in a pocket and thus is tradeable. Each unit of currency (e.g., a dollar) is identical and can be divided into fractions (e.g., cents) of the whole, making dollars effective units of account. A bill or coin, when well cared for, will not degrade substantively for years, meaning it can function as a store of value.

In the United States, paper currency and coins are produced by the Bureau of Engraving and Printing (BEP) and the United States Mint, respectively, both of which are units within the Department of the Treasury (Treasury). The Federal Reserve (the Fed) distributes the currency and coin to banks, savings associations, and credit unions upon request, and the banks in turn make the cash available to their customers. When a bank orders cash, the Fed deducts the amount from the bank's Fed account.[11] The revenues and costs to the government from this system are examined in the "Cash Effects on Government" section below.

Data suggests that the demand for cash in the United States has continued to grow despite the introduction of new payment services and systems. Fed data indicates that the amount of currency in circulation has increased steadily over at least the past 20 years (see Figure 2). As of December 31, 2018, there were more than 43 billion notes (more commonly called *bills*) worth over $1.67 trillion in circulation.[12] The Fed determines how many new notes "are needed to meet the public's demand

---

[10] Governor of the Riskbank Stefan Ingves, "Do We Need an E-Krona?" Speech to Swedish House of Finance, Stockholm, Sweden, August 12, 2017, pp. 3-4, at https://www.bis.org/review/r180123c.pdf.

[11] The Board of Governors of the Federal Reserve System, "Currency and Coin Services," at https://www.federalreserve.gov/paymentsystems/coin_about.htm.

[12] The Board of Governors of the Federal Reserve, "Currency and Coin Services: Data," at https://www.federalreserve.gov/paymentsystems/coin_data.htm. This is the most recent data available that includes both number and value of notes in circulation. More recent data available for the value of currency is used later in the report.

[, which]…reflects the Board's assessment of the expected growth rates for payments of currency to and receipts of currency from circulation."[13]

This growth in demand is not wholly surprising, because demand for cash would be expected to grow as does the economy, the population, and price levels. In addition, the demand for cash is growing because certain people may be increasingly using it solely as a store of value or safe investment (imagine the proverbial risk-averse saver keeping money under the mattress), rather than to make purchases.[14] In addition, there remains a high demand for U.S. currency abroad, both as a store of value and medium of exchange.[15]

Some evidence suggests people are using cash *for payments* less often. For example, according to preliminary findings of a Fed survey, cash transactions in the United States fell from 40.7% of all transactions in 2012 to 32.5% in 2015.[16] Taken together with data from the triennial *Federal Reserve Payment Study*, these survey results suggest the number of cash transactions during that time fell from roughly 84.8 billion per year to 69.4 billion.[17] However, Fed economists have subsequently noted that significant changes in the survey methodology and unaccounted for effects from economic conditions means the eight-point decline in the share of transactions "almost surely does not accurately reflect actual changes in consumer preferences for cash."[18] After making adjustments to account for

---

[13] The Board of Governors of the Federal Reserve, "Currency and Coin Services," at https://www.federalreserve.gov/paymentsystems/coin_about.htm.

[14] Morten Bech et al., "Payments Are A-Changin' But Cash Still Rules," Bank of International Settlements, *BIS Quarterly Review*, March 2018, pp. 67-73, 76-77, at https://www.bis.org/publ/qtrpdf/r_qt1803g.pdf.

[15] The Board of Governors of the Federal Reserve, "Currency and Coin Services," at https://www.federalreserve.gov/paymentsystems/coin_about.htm.

[16] Wendy Matheny, Shaun O'Brien, and Claire Wang, *The State of Cash: Preliminary Findings from the 2015 Diary of Consumer Payment Choice*, Federal Reserve System Cash Product Office, November 2016, pp. 1-4, at https://www.frbsf.org/cash/files/FedNotes-The-State-of-Cash-Preliminary-Findings-2015-Diary-of-Consumer- Payment-Choice.pdf.

[17] The Congressional Research Service's (CRS's) calculation using data from The Federal Reserve System, *The Federal Reserve Payments Study 2016*, December 2016, pp. 2-4, 12, at https://www.federalreserve.gov/newsevents/ press/other/2016-payments-study-20161222.pdf.

[18] Claire Greene, Shaun O'Brien, and Scott Schuh, *U.S. Consumer Cash Use, 2012 and 2015: Introduction to the Diary of Consumer Payment Choice*, Federal Reserve Bank of Boston, October 25, 2017, pp. 1-5, at https://www.bostonfed.org/publications/research-data-report/2017/us-consumer-cash-use-2012-and-2015.aspx.

these factors, those economists estimated the decline in the percentage of transactions that were in cash was roughly half of the initially estimated decline in the share of cash transactions.[19]

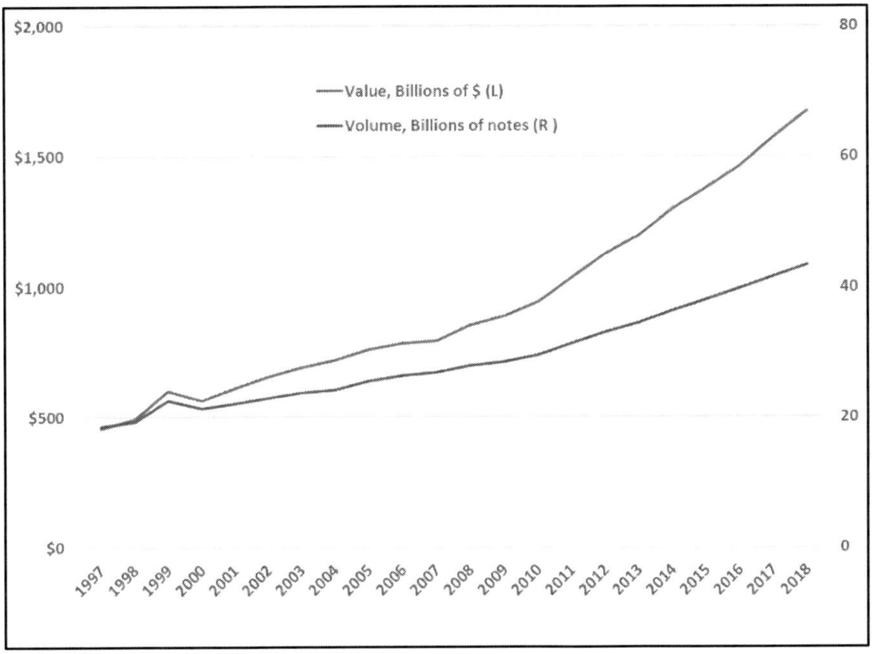

Source: The Federal Reserve Currency and Coin Services, at https://www.federalreserve.gov/paymentsystems/coin_data.htm, April 29, 2019.

Figure 2. U.S. Currency in Circulation.

The most recent data indicates that Americans used cash for 31% of their transactions in 2016, with stronger cash preference for small, in-person transactions (60% of in-person transactions under $10).[20]

---

[19] Ibid., p.32.

[20] Shaun O'Brien, *Understanding Consumer Cash Use: Preliminary Findings from the 2016 Diary of Consumer Payment Choice*, Federal Reserve Bank of San Francisco, Fednotes, November 28, 2017, at https://www.frbsf.org/cash/ publications/fed-notes/2017/november/ understanding-consumer-cash-use-preliminary-findings-2016-diary-of-consumer-payment-choice/.

Advantages and Costs of Cash

Cash Advantages for Consumers and Businesses

One of the main benefits of cash is that it is a simple, easy, robust payment mechanism that requires no ancillary technologies.[21] Payers and payees validate and settle transactions simply by physically exchanging the currency; the consumer needs no magnetic-stripe card or mobile device, and the seller does not need a card-reading machine or other payment-receiving device. Relatedly, some observers assert cash provides a security against potential disruptions to electronic payment systems. For example, in the event of a significant cyberattack or extended power outage, cash could continue to serve the functions of money while electronic payment systems could not.

Cash also acts as a safe asset in which to invest savings and its usage can involve a high degree of privacy, features that will be examined in more detail in the "Potential Costs and Risks of a Reduced Role for Cash" section below. In addition, holding cash might impart other psychological benefits to a consumer, such as feelings of greater control over budgeting and associations with wealth.[22]

Cash Costs for Consumers and Businesses

Using and accepting cash involves certain costs to consumers and businesses. For example, consumers may have to pay fees to withdraw cash from automatic teller machines (ATMs). Banks with more than $1 billion in assets are required to report their revenue from ATM fees, and a Congressional Research Service (CRS) analysis indicates those banks collected at least $1.9 billion in ATM fees in 2018.[23] Other costs—including consumer losses through theft, misplacement, or accidental destruction of cash—are more difficult to estimate.

---

[21] Ibid.

[22] Alan Wheatley, "Cash Is Dead, Long Live Cash," International Monetary Fund, *Finance & Development*, June 2017, at https://www.imf.org/external/pubs/ft/fandd/2017/06/ pdf/ wheatley.pdf.

[23] The figure is the sum of all positive values reported in banks' Consolidated Reports of Condition of Income, Schedule RI, Memorandum item 15.c.

Businesses must pay for cash management services, such as cash delivery with armored trucks (an industry with estimated annual U.S. revenues of $2.8 billion) and security systems to dissuade thieves or robbers from attempting to steal cash kept on the retailer's premises.[24] Despite these efforts, U.S. businesses lose about $40 billion in employee cash thefts per year.[25] Similar to consumer's costs, quantifying all the costs of cash to businesses presents challenges, as certain costs are not straightforward and easily measurable.[26] For example, some portion of retail staff and managers' paid time is spent counting cash and reconciling tills.

Cash Effects on Government Revenues

In addition to its impacts on consumers and businesses, cash directly affects government revenues through three main mechanisms: (1) *seigniorage* (i.e., the "profit" the government makes by producing cash), (2) Federal Reserve remittances to the Treasury, and (3) tax evasion. Two of these mechanisms—seigniorage and remittances—increase government revenues. The third mechanism—tax evasion, facilitated by the anonymous and difficult-to-trace nature of cash transactions—decreases government revenue.

Revenue Generating: Seignoriage

In general, the value of the physical currency produced by the government exceeds the cost incurred to produce it. For example, a $100 bill costs about 14 cents to print, generating revenues $99.86 greater than cost.[27] The profit generated by this difference is called seigniorage, and this income would decrease if demand for cash were to fall.[28] In FY2017, the

---

[24] IBIS World, *U.S. Armored Transportation Services, Industry at a Glance*, accessed on March 26, 2019, at https://clients1.ibisworld.com/reports/us/industry/ataglance.aspx?entid=5925 (subscription required).

[25] Bhaskar Chakravorti, "The Hidden Costs of Cash," *Harvard Business Review*, June 26, 2014, at https://hbr.org/2014/06/the-hidden-costs-of-cash.

[26] Fumiko Hayashi and William Keeton, "Measuring the Cost of Retail Payment Methods," *Federal Reserve Bank of Kansas City Economic Review*, Second Quarter 2012, pp. 37-42.

[27] Federal Reserve Board of Governors, "FAQ: How Much Does It Cost to Produce Currency and Coin," at https://www.federalreserve.gov/faqs/currency_12771.htm.

[28] Specifically, when the government produces physical money by minting coins or printing paper money, the expenses it incurs (for the material, machines, labor, etc.) are typically less than

U.S. Mint generated $391.5 million in net income from circulating coins[29] and the U.S. Bureau of Engraving and Printing generated revenues $693 million greater than expenses.[30]

Revenue Generating: Fed Remittances

The second source of cash-generating revenue is remittances, which are transferred from the Fed to the U.S. Treasury. Any income the Fed earns after expenses and dividends paid to member banks, it remits to the Treasury (in 2017, the amount was $80.6 billion), and hence becomes a source of revenue for the federal government. A significant expense for the Fed is the interest it pays on depository institutions' deposits held in their Fed accounts. Such payments accounted for $28.9 billion of the Fed's $35.4 billion total expenses in 2017.[31] However, currency is a Fed liability on which it pays no interest.[32] Recall that when a bank orders cash from the Fed, the Fed deducts the amount from the bank's account. Thus, the more cash that is in circulation, the less interest the Fed must pay, and the greater its remittances to Treasury. In January 2019, there was approximately $1.7 trillion of currency in circulation,[33] and the Fed (as of this publication) paid an annual interest rate of 2.4% on reserve balances.[34] By these measures, if all currency were instead bank reserve balances held at the Fed, it could

what the produced currency is worth. *Seigniorage* is the "profit" that the government collects by minting and printing currency.

[29] United States Mint, *2017 Annual Report*, p. 6, at https://www.usmint.gov/wordpress/wp-content/uploads/2018/01/2017-annual-report.pdf.

[30] Bureau of Engraving and Printing, *2017 Chief Financial Officer Performance and Accountability Report*, p. 5, at https://www.moneyfactory.gov/images/ 2017CFO_Report.pdf.

[31] Board of Governors of the Federal Reserve System, *104th Annual Report: 2017*, June 2018, pp. 101-102, at https://www.federalreserve.gov/publications/files/2017-annual-report.pdf.

[32] Banks and other depository institutions obtain currency to meet consumer demand from a Federal Reserve Bank. When the Reserve Bank sends the currency to the institution, it debits the institution's Federal Reserve account by the amount shipped. Thus, an increase in currency is matched by a reduction in reserve balances institutions hold in their Federal Reserve accounts. See The Board of Governors of the Federal Reserve, "Credit and Liquidity Programs and the Balance Sheet," at https://www.federalreserve.gov/ monetarypolicy/bst_frliabilities.htm.

[33] Board of Governors of the Federal Reserve System (US), Currency in Circulation [WCURCIR], Federal Reserve Bank of St. Louis, at https:// fred.stlouisfed.org/ series/WCURCIR, accessed on March 14, 2019.

[34] Data available for download at the Federal Reserve Board of Governors website at https://www.federalreserve.gov/datadownload/Choose.aspx?rel=PRates.

increase Fed expenses (and thus reduce government revenues) by more than $40 billion a year. If interest rates on reserves (which change when the Fed alters monetary policy) rose or fell, then expenses would increase or decrease, respectively, in this scenario.

Revenue Reducing: Tax Evasion

Because cash leaves no electronic record, wage earners and businesses are able to underreport (in general, illegally) how much cash they receive in order to reduce their tax payments. Thus, cash contributes to the *tax gap*—the difference between what the government is owed and what is actually paid. The most recent Internal Revenue Service estimate released in 2016 examined the tax gap for the years 2008-2010, and found that the gap due to underreporting averaged $387 billion a year.[35] This estimate does not directly measure how much underreporting is facilitated by cash payments, and the figure for recent years is likely to be different. However, it provides a general context for how much tax revenue the government does not collect due to underreporting that is at least in part made possible by cash transactions.

The Potential Decline of Cash Usage

Businesses have long set conditions under which they would not accept cash. For example, certain businesses refuse to accept high-denomination bills.[36] However, according to anecdotal reporting, retail businesses are increasingly deciding that the costs of transacting in cash are high enough that they would rather not accept it at all.[37] Notably, this is

---

[35] Alain Dubois et al., *Federal Tax Compliance Research: Tax Gap Estimates for Tax Years 2008-2010*, Internal Revenue Service, Publication 1415 (Rev. 5-2016), May 2016, pp. 6-9, at https://www.irs.gov/pub/irs-soi/p1415.pdf.

[36] For a legal analysis of this practice, see the inset text box, "The Legality of Refusing Cash."

[37] See Andy Newman, "Cash Might Be King, But They Don't Care," *New York Times*, December 25, 2017, at https://www.nytimes.com/2017/12/25/nyregion/no-cash-money-cashless-credit-debit-card.html; Katherine Bindley, "You Cash Is No Good Here. Literally," *Wall Street Journal*, December 28, 2018, at https://www.wsj.com/articles/ your-cash-is-no-good-here-literally-11546013696; and Paul Davidson, "More Retailers Go Cashless to Cut Costs, Theft as Holiday Shopping Ramps Up," *USA Today*, November 28, 2018, at https://www.usatoday.com/story/money/2018/11/28/holiday-shopping-more-retailers-just-saying-no-cash/2063747002/.

occurring at businesses at which transactions are typically in-person for small dollar amounts—traditionally viewed as the type of transactions for which cash is the least costly option. If these stories are in fact indicative of a sustained trend, widespread non-acceptance of cash could have a variety of effects on consumers, businesses, as well as society and the economy at large. One particular effect that has drawn significant attention, as well as litigation, is that non-acceptance of cash could potentially marginalize those that have limited access to the financial system or mobile technological devices. This issue is examined in the "Lack of Financial Access for Certain Groups" section later in the report.

The Legality of Refusing Cash[38]

Federal law provides that U.S. coins and currency "are legal tender for all debts, public charges, taxes, and dues."[39] Although cash is accordingly "a valid and legal offer of payment" for *pre-existing* debts, federal law does not require that private businesses accept cash as payment for goods and services.[40] Recently, an increasing number of restaurants and retailers have taken advantage of the absence of a federal requirement to accept cash by requiring their customers pay using credit or debit cards.[41] This trend has led some commentators to worry that such "no cash" policies disadvantage many low-income consumers who lack access to bank accounts and credit cards.[42] A number of states and localities have responded to this concern by enacting laws that require certain types of businesses to accept cash.[43] Accordingly, although federal law does not require businesses to accept cash, some states and localities require that certain types of businesses accept cash when engaging in certain types of transactions.

[38] This legal analysis has been provided by Jay Sykes, Legislative Attorney.

[39] 31 U.S.C. § 5103.

[40] *See* Miller v. Madison, No. 1:12-CV-0874, 2013 WL 2181240 at *5 (N.D.N.Y. May 20, 2013) (explaining that the federal legal-tender statute "applies only to payments to creditors on preexisting debts," and that "no federal statute mandat[es] that a private business, person, or organization must accept legal tender as payment for goods or services"); Legal Tender Status, U.S. DEP'T OF THE TREASURY, https://www.treasury.gov/resource-center/faqs/currency/pages/ legal-tender.aspx (last visited April 30, 2019).

[41] *See* Henry Grabar, *No Shirt, No Swipe, No Service*, SLATE (July 24, 2018), https://slate.com/business/2018/07/cashless-stores-and-restaurants-are-on-the-rise-to-the-delight-of-credit-card-companies.html.

[42] *See* Karen Zraick, *This Legislation Could Force Stores to Take Your Cash*, N.Y. TIMES (Feb. 20, 2019), https://www.nytimes.com/2019/02/20/business/cashless-payment.html.

[43] *See* City of Philadelphia, Pa., Bill No. 180943 (2019); N.J. STAT. ANN. § 56:8-2.33 (2019); MASS. GEN. LAWS ch. 255D § 10A (2019).

Traditional Noncash Payment Systems

Overview

Were cash to decline as a payment system, the most likely replacement—at least at the current time—would appear to be traditional noncash, electronic payment systems, such as debit cards, credit cards, or payment mobile apps.

In traditional noncash payment systems like those that are prevalent today, participants hold their money in an account at a bank or other financial intermediary that maintains accurate ledgers of how much money each customer has available. To make a payment, the payer instructs (using a physical check or an electronic message) the intermediary to transfer money to the recipient's account. If the recipient holds an account at a different intermediary, those intermediaries will send messages to each other via messaging networks connecting them, instructing each to make the necessary changes to their ledgers. The intermediaries validate the transaction, ensure the payer has sufficient funds for the payment, deduct the appropriate amount from the payer's account, and add that amount to the payee's account.[44] For example, in the United States, a retail consumer may initiate an electronic payment by swiping a debit card, at which time an electronic message is sent over a network instructing the purchaser's bank to send payment to the seller's bank. Those banks then make the appropriate changes to their account ledgers (possibly using the Fed's payment system) reflecting that value has been transferred from the purchaser's account to the seller's account.[45]

As with physical currency, digital entries in account ledgers can serve the three functions of money well for use in payments. Instructions to change entries in a ledger can easily be sent, making the values in the ledger easily tradable. Numerical entries can be denominated in identical and divisible units, making them good units of account. Because numbers

---

[44] Richard J. Sullivan, "The Federal Reserve's Reduced Role in Retail Payments: Implications for Efficiency and Risk," *Federal Reserve Bank of Kansas City Economic Review* (third quarter 2012), pp. 80–87. Hereinafter Sullivan, "Federal Reserve's Reduced Role."

[45] Ibid.

in a ledger can remain unchanged during periods when no transactions are made, they can serve as a store of value.

According to the most recent complete *Federal Reserve Payment Study* on noncash payments, the number of traditional noncash payments made in the U.S. totaled more than 144 billion transactions with a value of almost $178 trillion in 2015.[46] These included payments via debit cards (69.5 billion transactions worth $2.56 trillion), credit cards (33.8 billion transactions worth $3.16 trillion), automated clearing house payments (ACH; 23.5 billion transactions worth $26.83 trillion),[47] and check payments (17.3 billion payments worth $26.83 trillion). Between 2012 and 2015, the number of transactions of the three electronic systems, debit, credit, and ACH, grew at annual rates of 7.1%, 8.0%, and 4.9%, respectively. Their values grew by 6.8%, 7.4%, and 4.0%, respectively. Check payments declined by an annual rate of 4.4% by number and 0.5% by value. According to a recent supplement to that study, both the growth of electronic payments and the decline of check payments accelerated in 2017.[48]

Advantages and Costs of Traditional Noncash Payment Systems

Traditional Noncash Payment System's Advantages

Payment based on physically exchanging currency has some notable shortcomings that can be addressed by a payment system based on maintaining account ledgers. One is that physical currency requires both the payer and the payee to either (1) be physically near each other, allowing the physical currency to pass from the possession of the former

---

[46] Federal Reserve System, *The Federal Reserve Payments Study 2016*, December 2016, p. 12, at https://www.federalreserve.gov/newsevents/press/other/2016-payments-study-20161222.pdf.

[47] The automated clearinghouse (ACH) system is a nationwide network operated by the Federal Reserve regional banks and the private Electronic Payments Network through which depository institutions send each other electronic transfers. See The Board of Governors of the Federal Reserve System, "Automated Clearinghouse Services," at https://www.federalreserve.gov/paymentsystems/fedach_about.htm.

[48] National Automated Clearing House Association, "ACH Network Volume and Value Hit New Highs in 2017," press release, April 9, 2017, at https://www.nacha.org/news/ach-network-volume-and-value-hit-new-highs-2017.

into the possession of the latter; or (2) have a sufficient trust in each other that the payee believes an assurance that he or she will receive the currency later. Another shortcoming is that holders of physical currency may have little recourse if it is lost, stolen, or accidentally destroyed. If, instead, money is exchanged by making valid changes in ledgers maintained by trusted intermediaries, the exchange can be accomplished without the risk of lost, stolen, or damaged currency.[49]

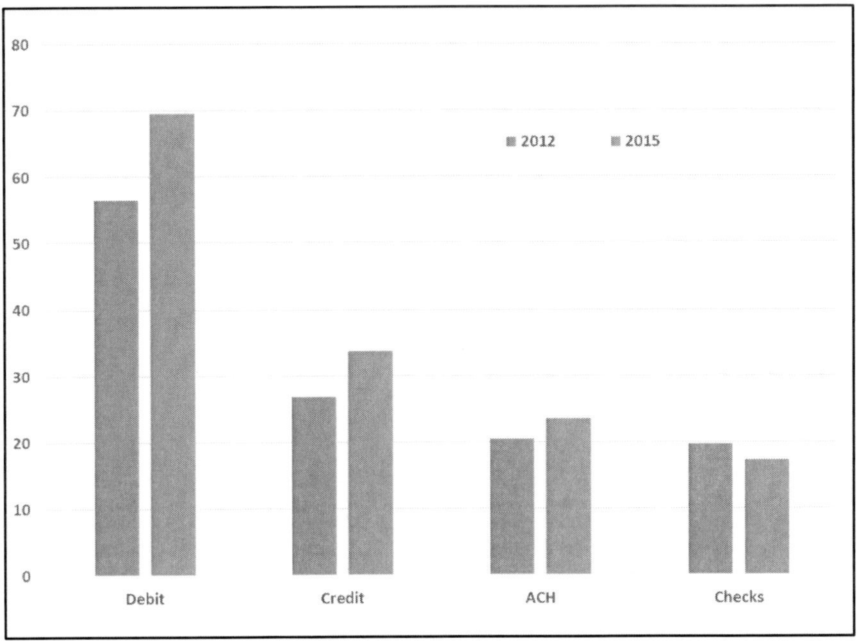

Source: Federal Reserve System, *The Federal Reserve Payments Study 2016*, December 2016, p. 12, at https://www.federalreserve.gov/newsevents/press/other/2016-payments-study-20161222.pdf.

Figure 3. Noncash Transactions, in Billions.

[49] Aleksander Berentsen and Fabian Schar, "The Case for Central Bank Electronic Money and the Non-case for Central Bank Cryptocurrencies," *Federal Reserve Bank of St. Louis Review*, vol. 100, no. 2 (second quarter 2018), p. 97-106, at https://files.stlouisfed.org/files/htdocs/publications/review/2018/04/16/the-case-for-central-bank-electronic-money-and- the-non-case-for-central-bank-cryptocurrencies.pdf.

In addition, noncash systems can make payments fast, easy, and convenient. Using them decreases the need for people to make regular estimations of how much cash they need to have on a particular day, the frequency of trips to the bank or ATM to get cash, and the amount of time waiting for cashiers to make change. Instead, a plastic card or an app on a mobile device can replace those activities with a card swipe or button push.

As information technology has progressed, the convenience has increased and the option to use electronic payments has become nearly ubiquitous. Until fairly recently, it was not uncommon for a retail establishment to reject card payments. Now, services such as Venmo, Apple Pay, and Google Pay, and card-reading devices, such as those made by Square, have made electronic payment options increasingly available, even for individuals to accept electronic payments from other individuals. The previously mentioned anecdotal reporting suggests there is a growing number of establishments that only accept electronic payments.[50]

For these systems to work well, participants must trust that banks and other intermediaries are keeping accurate ledgers that are changed only for valid transfers. Otherwise, an individual's money could be lost or stolen if a bank records the account as having an inaccurately low amount or transfers value without his or her permission. Another advantage of systems using traditional intermediaries is that they have a number of features that generate a high degree of trust and accuracy. Banks and other intermediaries have both a market and governmental incentive to be accurate. A bank or financial intermediary that does not have a good reputation for protecting a customer's money and processing transactions accurately would likely lose customers. In addition, governments typically subject banks to laws and regulations designed in part to ensure that banks are well run and that the money they hold is safe.[51] As such, banks take substantial measures to ensure security and accuracy.

---

[50] For example, see Andy Newman, "Cash Might Be King, But They Don't Care," *The New York Times*, December 25, 2017, at https://www.nytimes.com/2017/12/25/nyregion/no-cash-money-cashless-credit-debit-card.html.

[51] For more information on the regulation of banks in the United States, see CRS Report R44918, *Who Regulates Whom? An Overview of the U.S. Financial Regulatory Framework*, by Marc Labonte.

In addition, intermediaries generally are required to provide certain protections to consumers involved in electronic transactions, in part to protect them from losses resulting from unauthorized transfers. For example, the Electronic Fund Transfer Act (P.L. 95-630) limits consumers' liability for unauthorized transfers made using their accounts.[52] Similarly, the Fair Credit Billing Act (P.L. 93-495) requires credit card companies to take certain steps to correct billing errors, including when the goods or services a consumer purchased are not delivered as agreed.[53] Both laws also require financial institutions to make certain disclosures to consumers related to the costs and terms of using an institution's services.[54]

Traditional Noncash Payment System's Costs

Significant costs and physical infrastructure underlie systems for electronic money transfers to ensure the systems' integrity, performance, and availability. For example, payment system providers operate and maintain robust digital networks to connect retail locations with banks. The Fed operates and maintains electronics networks to connect banks to itself and each other.[55] These intermediaries store and protect huge amounts of data. Because these intermediaries are generally highly regulated, they incur regulatory compliance costs. Intermediaries recoup the costs associated with these systems and regulations and earn profits by charging fees directly when the system is used (such as the fees a merchant pays to have a card reading machine and "swipe fees" on each card transaction) or by charging fees for related services (such as checking account fees).

It is difficult to quantify how much traditional noncash payment systems cost and what portion of those costs is passed on to consumers and businesses. Performing a quantitative analysis is beyond the scope of this chapter.[56] What bears mentioning here is that certain costs of traditional

[52] 15 U.S.C. 1693g.
[53] 15 U.S.C. 1666.
[54] 15 U.S.C. 1693c and 15 U.S.C. 1632.
[55] Sullivan, "Federal Reserve's Reduced Role."
[56] Specifically, financial institutions offer an array of services, and the fees they charge are not always expressly linked to individual transactions. In addition, payment systems themselves differ in terms of services provided and costs incurred. Furthermore, analyzing the costs of this system in a way that is comparable to the cost structures of the cash services industry or

payment systems—and, in particular, the fees intermediaries in those systems charge—have at times been high enough to raise policymakers' concern and elicit policy responses.

For example, in response to businesses' assertions that Visa and MasterCard had exercised market power in setting debit card swipe fees at unfairly high levels, Congress included Section 1075 in the Dodd-Frank Consumer Protection and Wall Street Reform Act (Dodd-Frank Act; P.L. 111-203)—sometimes called the Durbin Amendment.[57] Section 1075 directs the Fed to limit debit card swipe fees charged by banks with assets of more than $10 billion.[58]

In addition, studies on unbanked and underbanked populations cite the fees associated with traditional bank accounts, a portion of which may be the result of providing payment services, as a possible cause for those populations' limited interaction with the traditional banking system.[59]

Although electronic payment systems protect customers from physical theft and are subject to a complex and sometimes overlapping array of state and federal laws, regulators, and regulations related to cybersecurity, they could nevertheless expose individuals to cyber-theft and identity theft. In addition, the systems themselves could be susceptible to disruption from cyberattacks.

The occurrence of successful hacks of banks and other financial institutions, wherein huge amounts of individuals' personal information are stolen or compromised, illustrates cyber-related risks. For example, in 2014, JPMorgan Chase, the largest U.S. bank, experienced a data breach that exposed financial records of 76 million households.[60] However, no

---

alternative payment systems discussed in the next section creates additional challenges. See Fumiko Hayashi and William R. Keeton, "Measuring the Costs of Retail Payment Methods," *Federal Reserve Bank of Kansas City Economic Review* (second quarter 2012), pp. 37-42, at https://www.kansascityfed.org/publicat/econrev/ pdf/12q2Hayashi-Keeton.pdf.

[57] For more information on debit card fees, see CRS Report R41913, *Regulation of Debit Interchange Fees*, by Darryl E. Getter.

[58] 15 U.S.C. 1693o-2.

[59] Susan Burhouse et al., *2015 FDIC National Survey of Unbanked and Underbanked Households*, FDIC, October 20, 2016, p. 3, at https://www.fdic.gov/householdsurvey/2015/2015report.pdf.

[60] Jessica Silver-Greenberg, Matthew Goldstein, and Nicole Perlroth, "JPMorgan Chase Hacking Affects 76 Million Households," *New York Times*, October 2, 2014, at https://dealbook.nytimes.com/2014/10/02/jpmorgan-discovers- further-cyber-security-issues.

consensus exists on how best to reduce the occurrence of such incidents, and whether current cybersecurity measures and regulatory frameworks are effective and efficient in mitigating cybersecurity risk is an open question.[61]

In addition, although the traditional electronic payment system is sufficiently fast and convenient to complete many transactions, other transactions can involve problematic delays. One such delay that can be particularly costly for consumers is the lag between when a payment (such as a paycheck) is deposited and when the full amount of the funds are available to the individual. Depending on factors related to which networks the payer's and payee's bank uses to process payments, it can take up to two business days (or more under certain circumstances) after a deposit is made for banks to fully validate, process, and settle the deposit.

Settlement delays can create a situation in which an individual has made a deposit that would give sufficient funds to pay a bill that is due, but nevertheless may overdraw the account because the deposit is awaiting processing. In such a situation, the individual faces a choice between costly outcomes—a late payment penalty on the bill, an overdraft fee on the bank account, or a fee from a check cashing or payday lending service. These costs are likely disproportionately borne by low- or moderate-income individuals who typically have low balances in their bank accounts. Faster or immediate payment processing could potentially reduce or eliminate costs incurred by individuals facing this situation.[62]

While delays in the payment system may seem anachronistic at a time when digital messages can be sent and data processed nearly instantaneously, the fact remains that aspects of the systems, networks, and infrastructures used today (including those operated by the Fed) were developed and deployed decades ago. Both the Fed and private institutions are working to increase system speed and efforts are underway to make

---

[61] For example, see Greg Baer and Rob Hunter, *A Tower of Babel: Cyber Regulation for Financial Services*, Bank Policy Institute, Banking Perspectives, second quarter 2017, at https://www.theclearinghouse.org/banking-perspectives/2017/2017-q2-banking-perspectives/articles/cyber-regulation-for-financial-services.

[62] Aaron Klein, *How the Fed Can Help Families Living Paycheck to Paycheck*, The Brookings Institution, November 22, 2017, at https://www.brookings.edu/research/how-the-fed-can-help-families-living-paycheck-to-paycheck/.

real-time payments in the United States the norm.[63] However, payment system operators arguably have little incentive to achieve faster or real-time payments because (1) they are in compliance with the current requirement facing banks pursuant to the Expedited Funds Availability Act of 1987 (P.L. 108-100) to generally make most types of deposits available by the second business day, (2) updating legacy systems is costly for the institutions that operate them, and (3) banks are generating revenue through overdraft fees.[64]

Alternative Electronic Payment Systems

Currently, it appears that the traditional noncash payment systems described above likely would replace cash payments should cash usage significantly decline. However some observers, citing the various costs and disadvantages associated with those systems—including delays in processing as well as reliance on traditional financial intermediaries—point to alternative electronic payment systems as potential dominant payment systems of the future.

Cryptocurrency, such as *Bitcoin*, is the most well-known of these alternatives. Described in more detail below, cryptocurrencies use *blockchain* technology and public or "distributed" ledgers to achieve validated transfers of digital representations of value.[65] The use of these

---

[63] See The Board of Governors of the Federal Reserve System, "Federal Reserve Board Seeks Public Comment on Potential Actions to Facilitate Real-time Interbank Settlement of Faster Payments," press release, October 3, 2018, at https://www.federalreserve.gov/ newsevents/ pressreleases/other20181003a.htm; and James Aramanda, *Real-Time Payments Have Arrived*, The Clearing House Banking Prespectives, Second Quarter 2018, at https:// www.theclearinghouse.org/banking-perspectives/2018/2018-q2-banking-perspectives/departments/real-time- payments-have-arrived.

[64] Aaron Klein, *The Fastest Way To Address Income Inequality? Implement a Real Time Payment System*, The Brookings Institution, January 2, 2019, at https://www.brookings.edu/ research/the-fastest-way-to-address-income-inequality-implement-a-real-time-payment-system/.

[65] For a more detailed examination of the *blockchain technology* underlying many of these systems, see CRS Report R45116, *Blockchain: Background and Policy Issues*, by Chris Jaikaran. For a more detailed examination of cryptocurrencies, see CRS Report R45427, *Cryptocurrency: The Economics of Money and Selected Policy Issues*, by David W. Perkins.

systems to make payments is quite rare relative to cash and traditional systems, and the role they will play in the future is speculative. Nevertheless, their potential to significantly affect the usage of cash and traditional systems for payments has drawn the attention of central banks. Some central banks are examining whether they should create a comparable payment system of digital currencies to offer the advantages of those systems themselves and to avoid being bypassed in the future.

This section briefly describes (1) existing private alternative electronic payment systems and (2) possible future central bank-run systems. With respect to alternative electronic payment systems, the section examines their potential advantages, costs, and obstacles to their widespread adoption. With respect to a potential central bank-run system, which is more speculative at this time, the section examines potential advantages and obstacles to their widespread adoption and uncertainties they present.

Private Payment Systems Using Distributed Ledgers

Overview

In general, private electronic payment systems using distributed ledgers allow individuals to establish an account identified by a string of numbers and characters (often called an *address* or *public key*) that is paired with a password or *private key* known only to the account holder.[66] A transaction occurs when two parties agree to transfer digital currency (perhaps in payment for a good or service) from one account to another. The buying party will "unlock" the currency used as payment with her private key, allowing some amount to be transferred from her account to the seller's. The seller then "locks" the currency in her account using her own private key.[67] From the perspective of the individuals using the system, the mechanics are similar to authorizing payment on any website

[66] In cryptography, a *key* is a value (e.g., a string of numbers) used for the operations of encryption, decryption, signature generation or signature verification.

[67] David Mills et al., *Distributed Ledger Technology in Payments, Clearing, and Settlement*, Board of Governors of the Federal Reserve System, Financial and Economics Discussion Series 2016-095, Washington, DC, 2016, pp. 10-14, at https://www.federalreserve.gov/econresdata/feds/2016/files/2016095pap.pdf.

that requires an individual to enter a username and password. In addition, companies offer applications or interfaces that users can download onto a device to make transacting in cryptocurrencies more user-friendly.

Many digital currency platforms use blockchain technology to validate changes to the ledgers.[68] In a blockchain-enabled system, payments are validated on a public or "distributed" ledger by a decentralized network of system users and cryptographic protocols.[69] In these systems, parties that otherwise do not know each other can exchange something of value (i.e., a digital currency) not because they trust each other but because they trust the platform and its protocols to prevent invalid changes to the ledger. A notable feature of transfers using blockchain is that they require no centralized, trusted intermediary such as a bank, government central bank, or other financial or government institution. Proponents envision these systems could achieve instantaneous transfers, although they currently require minutes or hours to finalize transfers.

The decentralized nature of digital currencies and their recent proliferation poses challenges to performing industry-wide analysis of their use in payments. For example, as of August 27, 2018, one industry group purported to track trading prices of 1,890 cryptocurrencies alone.[70] For brevity and clarity, this chapter uses statistics on Bitcoin—the first and most well-known cryptocurrency, the total value of which accounts for more than half of the industry as a whole[71]—as an illustrative example of a digital currency's use in payments.

In January 2017, the price of a Bitcoin on an exchange was about $993.[72] The price surged during the year, peaking at about $19,650 in

---

[68] For more information on blockchain technology, see CRS Report R45116, *Blockchain: Background and Policy Issues*, by Chris Jaikaran.

[69] Dylan Yaga et al., *Blockchain Technology Overview*, National Institute of Standards and Technology (NIST), NIST Interagency Report 8202, January 2018, pp. 12-25, at https://nvlpubs.nist.gov/nistpubs/ir/2018/NIST.IR.8202.pdf. Hereinafter Yaga, *Blockchain*.

[70] CoinMarketCap data, accessed at https://coinmarketcap.com/ on August 27, 2018.

[71] CoinMarketCap data, accessed at https://coinmarketcap.com/ on August 27, 2018; CRS calculation.

[72] Note on terminology: When discussing the exchange of one type of money for another, the term *exchange rate* is arguably more appropriate than the term *price*. However, in this instance, this chapter will follow popular convention and use the term *price*, as the notion that a Bitcoin or any other currency is purchased with dollars during an exchange is

December 2017, an almost 1,880% increase. However, the price then dramatically declined. Overall, the price of Bitcoin has experienced a high degree of volatility. On March 12, 2018, the price of Bitcoin was $3,860, down 80% from its peak. More recently, the price rebounded and was $5,948 on May 8, 2019.

Although price data on Bitcoin illustrates the public interest in and overall demand for this cryptocurrency, it is a poor indicator of how often it is being exchanged for goods and services (i.e., how often it is being used as money). Certain analyses appear to show that digital currencies are not being widely used and accepted as payment for goods and services, but rather as investment vehicles.[73]

The number of Bitcoin transactions may serve as a better indicator—though a flawed one—of the use of Bitcoin as a payment system. This number reveals how many times Bitcoins are transferred between accounts each day, and data indicates the number of transactions is miniscule compared with those of traditional systems. For example, in 2019 through March 12, the Bitcoin system averaged about 310,000 transactions per day globally, a pace that would result in about 113 million transactions per year.[74] Recall that in the United States alone, more than 144 billion traditional (nearly 1,275 times as many) noncash payments were made in 2015. Moreover, one problem with this measure it that it is a count of how many times two parties have exchanged Bitcoin, not a count of how many times Bitcoin has been used to buy something. Some portion of those exchanges, possibly a significantly large portion, is driven by investors giving fiat currency to an exchange to buy and hold the Bitcoin as an investment. In those transfers, Bitcoin is not acting as money (i.e., not being exchanged for a good or service).

---

essentially correct. Data retrieved from Federal Reserve Bank of St. Louis Economic Data, at https://fred.stlouisfed.org/series/CBBTCUSD accessed on May 9, 2019.

[73] Olga Kharif, "Bitcoin's Use In Commerce Keeps Falling Even as Volatility Eases," *Bloomberg*, August 1, 2018, at https://www.bloomberg.com/news/articles/2018-08-01/bitcoin-s-use-in-commerce-keeps-falling-even-as-volatility- eases.

[74] Data retrieved from bitcoin.com, https:// www.blockchain.com/ charts/ n-transactions?daysAverageString=7, on March 13, 2019.

Potential Advantages, Obstacles, and Costs to Private Payment Systems Using Distributed Ledgers

Advantages of Private Payment Systems Using Distributed Ledgers

As discussed in an earlier section, traditional electronic payment systems involve a number of intermediaries, such as government central banks and private financial institutions. To carry out transactions, these institutions operate and maintain extensive electronic networks and other infrastructure, employ workers, and require time to finalize transactions. To meet costs and earn profits, these institutions charge various user fees. Cryptocurrency advocates anticipate that a decentralized payment system operated through the internet could be less costly than traditional payment systems and existing infrastructures.[75] However, whether such efficiencies can or will be achieved remains an open question.

In addition, opening a bank account or otherwise using traditional electronic payment systems generally requires an individual to divulge to a financial institution certain basic personal information, such as name, Social Security number, and birthdate. Financial institutions store and may analyze or share this information. In some instances hackers have stolen personal information from financial institutions, causing concerns over how well these institutions can protect sensitive data.[76] Individuals seeking a higher degree of privacy or control over their personal data than that afforded by traditional systems may choose to use an alternative digital currency system that provides a degree of pseudonymity or anonymity.

Although inflation in the United States and other developed economies has been low in recent decades, some individuals may nevertheless believe that nontraditional digital money may maintain its value better than government-backed money in traditional systems. The dollar and most modern currencies are fiat money—that is, money that derives value based solely on government decree. Historically, incidents of hyperinflation in

---

[75] Jerry Brito and Andrea Castillo, *Bitcoin: A Primer for Policymakers* (Arlington, VA: Mercatus Center George Mason University, 2016), pp. 13-18, at https://www.mercatus.org/system/files/gmu_bitcoin_042516_webv3_0.pdf.

[76] For more information, see CRS Report R43496, *The Target and Other Financial Data Breaches: Frequently Asked Questions*, by N. Eric Weiss and Rena S. Miller.

certain countries have seen government-backed currencies lose most or nearly all of their value. Thus, some individuals may judge the probability of their fiat money losing a significant portion of its value to be undesirably high. These individuals may place greater trust in the ability of a decentralized network using cryptographic protocols that limit the creation of new money to maintain stable value of money than in the ability of government institutions to do so.[77]

Obstacles to Widespread Adoption of Private Payment Systems Using Distributed Ledgers

Several characteristics of cryptocurrency undermine its ability to serve the functions of money in a payment system. Currently, a relatively small number of businesses and individuals use or accept cryptocurrency for payment. As discussed above, Bitcoin transactions have averaged about 310,000 per day globally.[78] Cryptocurrency may be used as a medium of exchange less frequently than traditional money for several reasons. Unlike the dollar and most other government-backed currencies, cryptocurrencies are not legal tender, meaning creditors are not legally required to accept them to settle debts.[79] Consumers and businesses also may be hesitant to place their trust in these systems because they have limited understanding of them.[80] Relatedly, consumers and businesses may have sufficient trust in and be generally satisfied with traditional payment systems.

The recent high volatility in the price of many cryptocurrencies undermines their ability to serve as a unit of account and a store of value. Cryptocurrencies can have significant value fluctuations within short

---

[77] John O. McGinnis and Kyle W. Roche, "Why Bitcoin Is Booming," *Wall Street Journal*, July 9, 2017, https://www.wsj.com/articles/why-bitcoin-is-booming-1499638932.

[78] Data retrieved from bitcoin.com, at https:// www.blockchain.com/ charts/ n-transactions? daysAverageString=7, on March 13, 2019.

[79] Internal Revenue Service (IRS), "IRS Virtual Currency Guidance: Virtual Currency Is Treated as Property for U.S. Federal Tax Purposes; General Rules for Property Transactions Apply," Notice 2014-21, March 25, 2014, p. 1, https://www.irs.gov/pub/irs-drop/n-14-21.pdf. Hereinafter IRS, "Virtual Currency Guidance."

[80] Office of the Director of National Intelligence, *Risks and Vulnerabilities of Virtual Currency: Cryptocurrency as a Payment Method*, 2017 Public-Private Analytic Exchange Program, 2017, pp. 11-17, at https://www.dni.gov/files/PE/ Documents/9—2017-AEP_Risks-and-Vulnerabilities-of-Virtual-Currency.pdf.

periods of time; as a result, pricing goods and services in units of cryptocurrency would require frequent repricing and likely would cause confusion among buyers and sellers.[81]

Whether cryptocurrency systems are *scalable*—meaning their capacity can be increased in a cost- effective way without loss of functionality—is uncertain.[82] At present, the systems underlying cryptocurrencies do not appear capable of processing the number of transactions that would be required of a widely adopted, global payment system. One concern involves the significant energy consumption required to run and cool the computers that validate the transactions on these platforms.[83]

Costs of Private Payment Systems Using Distributed Ledgers

As the energy consumption of a digital currency system demonstrates, these systems are not costless. In addition to energy, they require computer hardware and facilities. Often making payments on these platforms involves paying fees. Whether these direct economic costs of using the system are fixed or—as they develop and mature—become less than existing systems is an open question.

Digital currency systems, at least as currently designed and regulated, also might impose other costs on society. Some critics of these systems fear their pseudonymous, decentralized nature may provide a new avenue for criminals to launder money, evade taxes, or sidestep financial sanctions. For example, Bitcoin was the currency used on the internet-based, illegal drug marketplace and Bitcoin escrow service called Silk Road. This marketplace facilitated more than 100,000 illegal drug sales from approximately January 2011 to October 2013, at which time the

---

[81] David Yemack, *Is Bitcoin a Real Currency? An Economic Appraisal*, National Bureau of Economic Research, Working Paper 19747, December 2013, pp. 2-3, at https://www.nber.org/papers/w19747.

[82] Kyle Croman et al., *On Scaling Decentralized Blockchains: A Position Paper*, International Conference on Financial Cryptography and Data Security, Christ Church, Barbados, February 26, 2016, at https://link.springer.com/content/pdf/10.1007%2F978-3-662-53357-4_8.pdf.

[83] Karl J. O'Dwyer and David Malone, "Bitcoin Mining and Its Energy Footprint," paper delivered at the 25th IET Irish Signals and Systems Conference, June 2014, pp. 280-285, at http://eprints.maynoothuniversity.ie/6009/1/DM- Bitcoin.pdf.

government shut down the website and arrested the individuals running the site.[84]

Consumer groups and other observers are also concerned that digital currency users are inadequately protected against unfair, deceptive, and abusive acts and practices. The way cryptocurrencies are sold, exchanged, or marketed can subject cryptocurrency exchanges or other cryptocurrency-related businesses to generally applicable consumer protection laws,[85] and certain state laws and regulations are being applied to cryptocurrency-related businesses.[86] However, other laws and regulations aimed at protecting consumers engaged in electronic financial transactions may not apply.

For example, the Electronic Fund Transfer Act of 1978 (EFTA; P.L. 95-630) requires traditional financial institutions engaging in electronic fund transfers to make certain disclosures about fees, correct errors when identified by the consumer, and limit consumer liability in the event of unauthorized transfers.[87] Because no bank or other centralized financial institution is involved in digital currency transactions, EFTA generally has not been applied to these transactions.[88] In addition, the laws and regulations that do apply generally have not been implemented specifically to address digital currencies or related businesses. Whether the current regulatory regime applied to digital currency transactions, but originally implemented for different financial activities (e.g., traditional money transmission), is effective and efficient is a debated issue.

Finally, some central bankers and other experts and observers have speculated that the widespread adoption of cryptocurrencies could affect

---

[84] The U.S. Attorney's Office Southern District of New York, "Ross Ulbricht, the Creator and Owner of the 'Silk Road' Website, Found Guilty in Manhattan Federal Court on All Counts," press release, February 5, 2015, at https://www.justice.gov/usao-sdny/pr/ross-ulbricht-creator-and-owner-silk-road-website-found-guilty-manhattan- federal-court.

[85] This chapter will focus on consumer protection laws. Potentially applicable federal or state civil and criminal fraud statutes are beyond the scope of this chapter.

[86] Nicholas Gess and Andrew Ray, "State Attorneys General to Fintech Companies: Eyes on Cryptocurrencies," in *All Things FinReg*, a blog of Lexology, July 31, 2018, at https://www.lexology.com/library/detail.aspx?g=baaab9f9-af12-49e6-99d5-b063b0e61533.

[87] 15 U.S.C §1693c, §1693f, §1693g.

[88] See CRS Report R43339, *Bitcoin: Questions, Answers, and Analysis of Legal Issues*, by Edward V. Murphy and M. Maureen Murphy.

the ability of the Fed and other central banks to implement and transmit monetary policy. The Fed conducts monetary policy with the goals of achieving price stability and low unemployment. Like other central banks it achieves its goals by, putting it simply, controlling the amount of money in circulation in the economy. If one or more additional currencies that the government did not control the supply of were also prevalent and viable payment options, it could limit central banks' ability to transmit monetary policy to financial markets and the real economy. In this scenario, central banks likely would have to make larger adjustments to the fiat currency they do control to have the same effect as previous adjustments. Another possibility is that they would have to start buying and selling the digital currencies themselves in an effort to affect the availability of these currencies.[89] These risks have led some central banks and other observers to suggest that perhaps central banks could issue their own digital currencies.[90]

Central Bank Digital Currencies

Overview

The risks and challenges posed by private digital currencies have led some observers to suggest that perhaps central banks should offer their own *central bank digital currencies* (CBDCs) to realize certain hoped-for efficiencies in the payment system in a way that would be "safe, robust, and convenient."[91] To date, no country has successfully created a CBDC for payment use by the general public.[92]

---

[89] Dong He, "Monetary Policy in the Digital Age," *International Monetary Fund: Finance and Development*, vol. 55, no. 2 (June 2018), p. 15, at https://www.imf.org/external/pubs/ft/fandd/2018/06/central-bank-monetary-policy-and- cryptocurrencies/he.pdf.

[90] Bank for International Settlements, Committee on Payments and Market Infrastructures, *Central Bank Digital Currencies*, March 12, 2018, pp. 1-2, at https://www.bis.org/cpmi/publ/d174.htm. Hereinafter Bank for International Settlements, *Central Bank Digital Currencies*.

[91] Bank for International Settlements, *Central Bank Digital Currencies*.

[92] Venezuela has developed government-backed cryptocurrencies in apparent attempts to evade U.S. financial sanctions. However, their viability as a payment system has faced numerous hurdles. See Brian Ellsworth, "Special Report: In Venezuela, New Cryptocurrency is Nowhere to Be Found," *Reuter*, August 30, 2018, at https://www.reuters.com/article/us-

The extent to which a central bank could or would want to create a new, digital-only payment system likely would be weighed against the consideration that these government institutions already have trusted digital payment systems in place. Because of such considerations, the exact form that CBDCs would take could vary across a number of features and characteristics.[93] Nevertheless, some central banks are examining the idea of CBDCs and the possible benefits and issues they may present.[94]

For the purposes of this discussion, this chapter examines a CBDC that would be available to consumers for retail payments. Some proposals would limit CBDCs to wholesale payments between banks and other financial institutions.

Potential Advantages, Obstacles, and Uncertainties of CBDCs

Potential Advantages of CBDCs

Proponents of CBDCs generally argue they could provide efficiency gains over traditional legacy systems and contend that central banks could use the technologies underlying digital currencies to deploy a faster, less costly government-supported payment system.[95]

Observers have speculated that a CBDC could take the form of a central bank allowing individuals to hold accounts directly at the central bank. Advocates argue that a CBDC created in this way could increase systemic stability by imposing additional discipline on commercial banks. Because consumers would have the alternative of safe deposits made directly with the central bank, commercial banks likely would have to offer interest rates and limit risks at levels necessary to attract deposits above any deposit insurance limit.[96]

---

cryptocurrency-venezuela-specialrepor/special-report-in-venezuela-new- cryptocurrency-is-nowhere-to-be-found-idUSKCN1LF15U.

[93] Bank for International Settlements, *Central Bank Digital Currencies*, pp. 3-10.

[94] For example, see Bank for International Settlements, *Central Bank Digital Currencies*.

[95] Eswar Prasad, *Central Banking in a Digital Age: Stock-Taking and Preliminary Thoughts*, Hutchins Center on Fiscal and Monetary Policy at Brookings, April 2018, p. 14, at https://www.brookings.edu/wp-content/uploads/2018/04/es_20180416_digitalcurrencies.pdf. Hereinafter Prasad, *Central Banking*.

[96] Aleksander Berensten and Fabian Schar, "The Case for Central Bank Electronic Money and the Non-case for Central Bank Cryptocurrencies," *Federal Reserve Bank of St. Louis Review*,

In addition, CBDCs could increase government revenue through a seignoriage-like mechanism. A more expansive definition of seignoriage is the income government obtains from having government liabilities act as money. Physical money—because it is liquid and low-risk—earns no interest rate and carries a cost to produce. Money—both physical and electronic in the traditional system—is also a balance sheet liability to the issuing authority, such as the Fed or other central banks.[97] If the Fed allowed individuals to hold accounts directly with the Fed, the Fed would issue low- or no-interest liabilities to individuals (as electronic entries in a ledger produced at less cost than physical currency). Then, as happens now, the Fed would use those liabilities to fund purchases of assets that earn a higher interest rate than what the Fed pays on liabilities. This would produce income, perhaps greater income than is earned through traditional seignoriage.

Potential Obstacles to Creation of CBDCs

One of the main arguments critics—including various central bank officials—make against CBDCs is that there is no "compelling demonstrated need" for such a currency, because central banks and private banks already operate trusted electronic payment systems that generally offer fast, easy, and inexpensive transfers of value.[98]

Opponents also argue that a CBDC in the form of individual direct accounts at the central bank would reduce the role of private banks in financial intermediation and potentially expand the role of government central banks inappropriately. A portion of consumers likely would shift their deposits away from private banks toward central bank digital money, which would be a safe, government-backed liquid asset. Deprived of this

---

vol. 100, no. 2 (second quarter 2018), at https://research.stlouisfed.org/ publications/review/2018/02/13/the-case-for-central-bank-electronic-money-and-the-non-case-for-central-bank-cryptocurrencies/.

[97] Federal Reserve Chairman Alan Greenspan, "Regulation of Electronic Payment Systems," Remarks at the U.S. Treasury Conference on Electronic Money and Banking: The Role of Government, Washington, DC, September 19, 1996, at https://www.federalreserve.gov/boarddocs/speeches/1996/19960919.htm.

[98] Lael Brainard, *Cryptocurrencies, Digital Currencies, and Distributed Ledger Technologies: What Are We Learning?* Board of Governors of the Federal Reserve System, May 15, 2018, at https://www.federalreserve.gov/newsevents/speech/files/brainard20180515a.pdf.

funding, private banks likely would have to reduce their lending, leaving central banks to decide whether or how they should support lending markets to avoid a reduction in credit availability.[99]

In addition, skeptics of CBDCs object to the assertion that these currencies would increase systemic stability, arguing that CBDCs would create a less-stable system because they would facilitate runs on private banks. These critics argue that at the first signs of distress at an individual institution or the bank industry, depositors would transfer their funds to this alternative liquid, government-backed asset.[100]

Uncertainty: CBDCs' Potential Effects on Monetary Policy

Observers also disagree over whether CBDCs would have a desirable effect on central banks' role and abilities in carrying out monetary policy. Proponents argue that, if individuals held a CBDC on which the central bank set interest rates, the central bank could directly transmit a policy rate to the macroeconomy, rather than achieving transmission through the rates the central bank charges banks and the indirect influence of rates in particular markets.[101] In addition, if holding cash (which in effect has a 0% interest rate) were not an option for consumers, central banks potentially would be less constrained by the *zero lower bound*.[102] The zero lower bound is the idea that the ability of individuals and businesses to hold cash and thus avoid negative interest rates limits central banks' ability to transmit negative interest rates to the economy.

Critics argue that taking on such a direct and influential role in private financial markets is an inappropriately expansive role for a central bank. They assert that if CBDCs were to displace cash and private bank deposits, central banks would have to increase asset holdings, support lending markets, and otherwise provide a number of credit intermediation activities

---

[99] Bank for International Settlements, *Central Bank Digital Currencies,* p. 9.

[100] Bank for International Settlements, *Central Bank Digital Currencies*, p. 16.

[101] Bank for International Settlements, *Central Bank Digital Currencies*, pp. 11-12.

[102] Eswar Prasad, *Central Banking in a Digital Age: Stock-Taking and Preliminary Thoughts*, Hutchins Center on Fiscal and Monetary Policy at Brookings, April 2018, pp. 19-20, at https://www.brookings.edu/wp-content/uploads/2018/04/es_20180416_ digitalcurrencies.pdf.

that private institutions currently perform in response to market conditions.[103]

Part II: Potential Implications of a Reduced Role for Cash

As discussed above, although cash remains a frequently used payment system, other payment systems continue to develop that offer their own advantages and costs. Various trends suggest that due to market preference or government policy, the role of cash in the payment system has begun to decline and may continue to decline, perhaps significantly, in coming years. If the relative benefits and costs of cash and the various other payment methods evolve in such a way that cash is significantly displaced as a commonly accepted form of payment, that evolution could have a number of effects, both positive and negative, on the economy and society. This section of the report describes a number of potential benefits of a reduced role for cash in the U.S. economy and the various risks and costs that may occur. Many of the factors discussed below may not occur wholly as a benefit, risk, or cost; rather, a potential benefit may bring with it a risk, and vice versa.

Potential Benefits of a Reduced Role for Cash

Some observers argue that reducing or eliminating cash payments in the U.S. economy will produce certain beneficial outcomes, including improved efficiency in payments, reduced criminality, and improved ability for the Fed to implement certain monetary policy. As discussed below, although these outcomes generally may be beneficial, that does not mean that there are not certain costs, or drawbacks, that may counterbalance these positive effects.

---

[103] Bank for International Settlements, *Central Bank Digital Currencies,* p. 14.

More Efficient Payments

Proponents of noncash payment systems assert that net economic benefits from the use and maintenance of a cash payment system are (or will become as technology advances) less than the net benefits of using and maintaining noncash systems. Put another way, they argue that the resources, labor, and capital that go into the cash system—for example, producing currency; stocking and maintaining ATMs; safely transporting cash; protecting businesses from theft and robbery—make it less efficient than noncash systems.[104] If true—and absent policy interventions—market forces likely will result in the displacement of cash by other payment methods as businesses increasingly choose not to accept cash and consumers increasingly prefer not to use it.

Under this scenario, although the payment system on net may be more efficient, it would not necessarily be true that all people would benefit, as is discussed in the "Potential Costs and Risks of a Reduced Role for Cash" section.

Impediment to Crime

Proponents of cashless societies assert that the elimination of cash would reduce crime by making operating an illegal enterprise more difficult and certain crimes, such as robbery and burglary, less remunerative.[105] These proponents argue that criminals are more likely to conduct business in cash and to hold cash as an asset, in large part because cash is anonymous and allows them to avoid establishing relationships with and generating records at financial institutions that may be subject to anti-money laundering reporting and compliance requirements.[106] Accordingly, they assert that the elimination of cash would be beneficial on net, because operating a criminal enterprise would become more difficult.

---

[104] Patrick Jenkins, "'We Don't Take Cash': Is This the Future of Money," *The Financial Times*, May 10, 2018, at https://www.ft.com/content/9fc55dda-5316-11e8-b24e-cad6aa67e23e.

[105] Kenneth S. Rogoff, "Costs and Benefits to Phasing Out Paper Currency," pp. 2-6.

[106] For example, banks and other financial institutions are subject to Bank Secrecy Act reporting and compliance requirements (31 U.S.C. §§ 5311-5332).

Certain studies have shown that the prevalence of cash is correlated with the incidence of crime.[107] In addition, the amount of "strong" currencies (i.e., highly valuable and highly stable currencies) in circulation exceeds what many people would consider a reasonable amount needed for typical consumer transactions. For example, with the U.S. population at approximately 329 million,[108] the $1.6 trillion of currency in circulation equates to about $4,900 per person. Proponents of a cashless society assert that this number is inflated due in part to the cash demand of criminals (although part is also due to demand for the U.S. dollar from abroad).[109]

Although a robust analysis of this question is beyond the scope of this chapter, arguments that cash facilitates crime and even that reducing cash may reduce crime appear in certain cases to be well founded.[110] However, when analyzing the net benefit to society of going cashless, reduced crime should be weighed against any cost that a reduction in cash would impose on legitimate cash users. One such legitimate group is examined in more detail in the "Lack of Financial Access for Certain Groups" section below. The effect a reduction in cash payments would have on crime should not be overstated, as criminals likely would seek other ways to commit and hide their crimes. For example, the prevalence of cybercrime may increase.

Elimination of a Monetary Policy Constraint

Another benefit (from a macroeconomic perspective) of a cashless society cited by economists would be the potential elimination of the practical inability of central banks, such as the Fed, to implement negative interest rates. When an economy is in recession or otherwise performing poorly, one monetary policy response is to lower interest rates. Lower interest rates can spur companies to borrow in order to invest and spur

---

[107] For example, see Richard Wright et al., *Less Cash, Less Crime: Evidence from the Electronic Benefit Transfer Program*, National Bureau of Economic Research, Working Paper no. 19996, March 2014, at https://www.nber.org/ papers/w19996.pdf. Hereinafter Richard Wright et al., *Less Cash, Less Crime: Evidence from the Electronic Benefit Transfer Program*.

[108] U.S. Census Population Clock, at https://www.census.gov/popclock/, accessed on May 9, 2019.

[109] Kenneth S. Rogoff, "Costs and Benefits to Phasing Out Paper Currency," pp. 3-4.

[110] Richard Wright et al., *Less Cash, Less Crime: Evidence from the Electronic Benefit Transfer Program*.

consumers to borrow in order to make additional purchases, thus boosting economic activity and mitigating the impact of recessions. However, many economists believe that policymakers are constrained by a zero lower bound—that whatever policy rate they may set, interest rates in many markets will not fall below zero.[111] The reason is that holding cash offers a zero interest rate. Thus, if the Fed attempted to implement negative interest rates, individuals could avoid those rates by transferring their funds into cash. If holding cash was not an available option, it would be easier for negative interest rates to be transmitted to more financial markets.

However, any benefit provided by increasing policymakers' ability to affect the macroeconomy with negative interest rates should be weighed against the cost it would impose on the individual savers whose account balances would decrease in value during a period of negative interest rates.

Potential Costs and Risks of a Reduced Role for Cash

Skeptics of reducing or eliminating the role of cash in the economy assert that cash serves a number of beneficial purposes, and argue that eliminating it would have adverse effects on certain financially vulnerable groups, eliminate an asset that provides safety against cyber vulnerabilities and financial crises, and reduce individuals' privacy. As with potential benefits to a reduction in cash, many of the factors discussed below may not occur wholly as a risk or cost, and they must be weighed against potential benefits when considering their overall impact.

---

[111] See Ben Bernanke, *How Big a Problem Is the Zero Lower Bound on Interest Rates*, Brookings Institution, Washington, DC, April 12, 2017, at https://www.brookings.edu/blog/ben-bernanke/2017/04/12/how-big-a-problem-is-the-zero-lower-bound-on-interest-rates/. Although central banks are constrained in their ability to implement negative rates, they are not incapable of doing so. For example, the European Central Bank and the Swiss National Bank (the Swiss central bank) have recently implemented negative policy rates (generally the rates banks pay to hold deposits at the central bank). It is believed that the availability of cash may limit how low rates can go, and how low interest rates for other holdings besides central bank reserves can go.

Lack of Financial Access for Certain Groups

If the United States were to move toward becoming a cashless society that required consumers to use noncash, electronic payment services, it could present difficulties for those segments of the population who lack access to the financial system or to an electronic network. Access to electronic payments typically requires an account with some financial institution, usually a bank. Often—and increasingly—it also involves using or accessing a device connected to the internet. However, these factors can present hardships and obstacles for certain vulnerable groups. The Federal Deposit Insurance Corporation reported that in the United States in 2015, 9 million households were unbanked, meaning that no member had a bank account. Of these, 37.8% reported that the main reason was that they do not have enough money to keep in an account, 9.4% reported that fees were too high, and 1.9% reported fees were unpredictable.[112] In total, this indicates almost half of the total unbanked, or roughly 4.5 million households, do not access banking services due to economic obstacles.

Sweden has been at the forefront of the move away from cash (see Appendix), and observers there, including Stefan Ingves, governor of Sweden's central bank, have voiced some of these concerns about going cashless.[113] In addition, anecdotal reporting indicates that retirees in Sweden are finding the change difficult and costly.[114]

In the United States, many assert that it would be beneficial to bring the unbanked into the banking system. Nevertheless, if the unbanked engaged with the banking system at a relatively high cost only because cash (which was a less expensive option for them) was no longer available, it would likely be a detrimental outcome for this group. Conversely, if the move to a cashless system led to less costly financial access for this group, they may stand to benefit.

---

[112] Federal Deposit Insurance Corporation, *FDIC Survey of Unbanked and Underbanked Households: Executive Summary*, October 20, 2016, pp. 1-3, at https://www.fdic.gov/householdsurvey/2015/2015execsumm.pdf.

[113] See Stefan Ingves, "Going Cashless," *International Monetary Fund, Finance and Development*, June 2018, at https://www.imf.org/external/pubs/ft/fandd/2018/06/central-banks-and-digital-currencies/point.htm.

[114] Maddy Savage, "The Swedes Rebelling Against a Cashless Society," *BBC News*, April 6, 2018, pp. at https://www.bbc.com/news/business-43645676.

Loss of Safety Provided by Cash

Proponents of cash often cite the robustness of physical currency as a payment system. Once in an individual's possession, cash does not rely on financial institutions or information technology (IT) based payment networks.[115] These proponents argue that if payments became entirely electronic, events such as power outages, hacker attacks, or (in the event of future cyber war) a state-sponsored attack would be capable of shutting down the most simple financial transaction— the exchange of money for goods and services. The financial system is already exposed to these threats to varying degrees, but the argument is that the elimination of cash amplifies those risks.

Because it functions well as a store of value, cash is a relatively safe asset in which to invest savings with no risk of losses resulting from a decline in a securities value or the failure of financial institutions or other entities.[116] The perceived safety of cash and its non-reliance on financial institutions also makes it desirable in times of financial turmoil or distress, when confidence in such institutions decreases. During these periods, many people prefer assets that are free from credit risk. For some of these individuals, deposit insurance guarantees may not wholly eliminate their fear of losses, whereas the safety of physical currency would.[117] Holding cash, then, could also provide a sense of security to risk-adverse people that may mistrust the financial system.

Privacy Concerns

Opening a bank account or otherwise using traditional noncash payment systems generally requires an individual to divulge certain basic personal information, such as name, Social Security number, and birthdate, to a financial institution. Financial institutions store this information and

---

[115] Laurens Cerulus and Cat Contiguglia, "Central Bankers Warn of Chaos in a Cashless Society," *Politico*, August 14, 2018.

[116] Morten Bech et al., "Payments Are A-Changin' But Cash Still Rules," Bank of International Settlements, *BIS Quarterly Review*, March 2018, pp. 67-73, 76-77, at https://www.bis.org/publ/qtrpdf/r_qt1803g.pdf.

[117] Governor of the Riskbank Stefan Ingves, "Do We Need an E-Krona?" Speech to Swedish House of Finance, Stockholm, Sweden, August 12, 2017, p. 5 at https://www.bis.org/review/r180123c.pdf.

information about the transactions linked to this identity. Under certain circumstances, they may analyze or share this information, such as with a credit-reporting agency. In some instances hackers have stolen personal information from financial institutions, causing concerns over how well these institutions can protect sensitive data. Finally, provided it follows proper legal procedures, the government also can access this information under certain circumstances.[118] Similarly, although new alternative payment systems may offer a degree of anonymity or pseudonymity, these systems still generate an unalterable record of transactions between parties.

Cash, by contrast, can be used anonymously, and people may wish to use cash for legitimate purposes to ensure their privacy. Certain consumers who are uncomfortable divulging and generating private information—even basic information that a transaction occurred—may prefer cash to any electronic payment methods.[119]

PROSPECTUS

Cash has a number of advantageous features that has made it a simple and robust payment system throughout most of human history. It is difficult to imagine conditions under which cash would be replaced entirely, and disappear from the economy, at least in the near future. Nevertheless, its hegemony as a payment system appears to have come to an end, as electronic payment systems have gained popularity, and the ubiquity of cash acceptance for in-person purchases also seems precarious. If noncash payment systems significantly displace cash and cash usage and acceptance significantly declines, there would be a number of effects (both positive and negative) on the economy and society. Now or in the near

---

[118] Catherine Downey, "The High Price of a Cashless Society: Exchanging Privacy Rights for Digital Cash," *The John Marshall Journal of Information Technology and Privacy Law*, vol. 14, no. 2 (Winter 1996), pp. 303-315.

[119] For example, see Matthew Campbell, "Germany Is Still Obsessed With Cash," *Bloomberg*, February 6, 2018, at https://www.bloomberg.com/news/features/2018-02-06/germany-is-still-obsessed-with-cash.

future, policymakers may face decisions about whether to impede or hasten the decline of cash and consider the implications of doing so.

APPENDIX. INTERNATIONAL CASE STUDIES

Two countries provide interesting case studies of market forces drastically changing the way a society makes payments.

Sweden

In recent years, the use of cash in Sweden has quickly and substantially declined, dropping from 40% to 13% of transactions between 2010 and 2018.[120] In many cases, businesses no longer accept cash, and one survey indicated that two-thirds of small businesses planned to stop accepting cash. Anecdotal reporting indicates that about 5% of bank branches accept cash deposits or offer cash withdrawals.[121] Furthermore, Sweden's central bank is examining the possibility of creating registered accounts for the purpose of issuing currency electronically.[122] Reportedly, many Swedes are generally in favor of the trend (the displacement of cash is due largely to consumer preference), though some have voiced concerns about financial access issues that the change causes for certain groups, such as the elderly.[123]

Observers have put forward a number of explanations for the Sweden's growing preference for electronic payment methods such as cards and

[120] Sveriges Riksbank (The Swedish Central Bank), *Payment Patterns in Sweden 2018*, May 2018, pp. 5-6, https://www.riksbank.se/globalassets/media/statistik/betalningsstatistik/2018/ payments-patterns-in-sweden-2018.pdf.

[121] Patrick Jenkins, "'We Don't Take Cash': Is This the Future of Money?" *The Financial Times*, May 10, 2018, at https://www.ft.com/content/9fc55dda-5316-11e8-b24e-cad6aa67e23e.

[122] Sveriges Riksbank (The Swedish central bank), *The E-Krona Project First Interim report*, September 2017, at https://www.riksbank.se/globalassets/media/rapporter/e-krona/2017/ rapport_ekrona_uppdaterad_170920_eng.pdf.

[123] Governor of the Riskbank Stefan Ingves, "Do We Need an E-Krona?" Speech to Swedish House of Finance, Stockholm, Sweden, August 12, 2017, p. 5, at https:// www.bis.org/ review/r180123c.pdf.

mobile app enabled payments. One argument asserts that Sweden is an especially technology savvy country. As such, Swedes are comfortable using electronic payment systems, and Swedish companies have developed fast and easy payment technologies, such as iZettle and Swish. Some observers also have suggested that Swedes are especially trusting of institutions and thus have fewer privacy concerns. Some have noted that the timing of the start of the decline in cash use among Swedes coincided with the start of a transition to new Swedish banknotes and coins. They suggest that this spurred people and businesses to make a switch not to the new bills and coins but instead to electronic payment methods.[124]

Kenya

In 2007, a company named Safaricom—Kenya's largest mobile phone network operator— introduced a "mobile money" service called M-Pesa ("M" stands for "mobile" and "pesa" is the Swahili word for money). Users of the service download a phone application and deposit cash with M-Pesa employees called "agents." They can then transfer money into any other M-Pesa account using their phone. Originally intended as a service for Kenyans who had moved to a city to earn money to send back home to rural areas, the service became tremendously popular as a general use payment system.[125] By 2016, there were approximately 31.6 million mobile money accounts in Kenya,[126] which had a total population of 47.6 million in 2017.[127]

---

[124] Maddy Savage, "Why Sweden is Close to Becoming a Cashless Economy," *BBC News*, September 17, 2017, at https://www.bbc.com/news/business-41095004.

[125] Alliance for Financial Inclusion, *Enabling Mobile Money Transfer: The Central Bank of Kenya's Treatment of M- Pesa*, 2010, at https://www.gsma.com/mobilefordevelopment/wp-content/uploads/2013/09/ enablingmobilemoneytransfer92.pdf. Hereinafter Alliance for Financial Inclusion, *Enabling Mobile Money Transfer*.

[126] William Cook and Claudia McKay, *Banking in the M-Pesa Age*, The Consultative Group to Assist the Poor, September 2017, p. 2, at https://www.cgap.org/research/ publication/ banking-m-pesa-age.

[127] Central Intelligence Agency, World Factbook, at https://www.cia.gov/library/publications/the-world-factbook/geos/ke.html.

Many observers identify the combination of lack of access to traditional banking services and the proliferation of mobile phones in Kenya as a driving factor for the expansion of M-Pesa and subsequent mobile money services. These observers argue that in Kenya, as with many developing and largely rural nations, both consumers and banks view financial and bank services as a business need of the rich. In 2006, before the introduction of M-Pesa, just 19% of Kenyans had bank accounts and there was 1.5 bank branches for every 100,000 people. However, 54% of Kenyans had their own mobile phone or access to one.[128]

Another explanation for the rise of mobile money is that Safaricom successfully identified a large, profitable, and previously untapped market in Kenya. Available mobile technology and its proliferation among the population meant low-cost money transfers could be profitably offered to lower-income consumers.[129] Certain observers assert that the success of M-Pesa has caused Kenyan financial institutions to reevaluate their business models, shifting their focus to offering services to lower-income groups than they previously targeted, and cite the increase in bank accounts and the decline of the average account balances as evidence of this change. As a result, the portion of the Kenyan population with access to some type of formal financial services has grown from 27% in 2006 to 75% in 2017.[130]

Although mobile money appears to have filled a market need, the degree to which it has displaced cash should not be overstated. An official at Safaricom estimated in February 2018 that as many as 8 out of 10 transactions are still cash transactions, as Kenyans still reportedly prefer cash for small, in-person purchases because of convenience and using M-

---

[128] Alliance for Financial Inclusion, *Enabling Mobile Money Transfer*.

[129] "Why Does Kenya Lead the World in Mobile Money," *The Economist*, March 2, 2015, at https://www.economist.com/the-economist-explains/2015/03/02/why-does-kenya-lead-the-world-in-mobile-money.

[130] William Cook and Claudia McKay, *Banking in the M-Pesa Age*, The Consultative Group to Assist the Poor, September 2017, p. 2, at https://www.cgap.org/research/publication/banking-m-pesa-age.

Pesa generally involves fees.[131] In addition, workers are still generally paid in cash.[132]

131 Zoe Flood, "Zimbabwe and Kenya Lead the Way in Africa's Dash From Cash," *The Guardian*, February 22, 2018, at https://www.theguardian.com/world/2018/feb/22/kenya-leads-way-mobile-money-africa-shifts-towards-cash-free- living.

132 Ibid.

INDEX

A

B

D

E

F

G

H

I

J

L

M

R

S

T

U